AF676659

Leptospira infections in Domestic and Wild Animal

Leptospira infections in Domestic and Wild Animal

Editors

Giovanni Cilia
Fabrizio Bertelloni
Filippo Fratini

MDPI • Basel • Beijing • Wuhan • Barcelona • Belgrade • Manchester • Tokyo • Cluj • Tianjin

Editors

Giovanni Cilia
Department of Veterinary Sciences, University of Pisa
Italy

Fabrizio Bertelloni
Department of Veterinary Sciences, University of Pisa
Italy

Filippo Fratini
Department of Veterinary Sciences, University of Pisa
Italy

Editorial Office
MDPI
St. Alban-Anlage 66
4052 Basel, Switzerland

This is a reprint of articles from the Special Issue published online in the open access journal *Pathogens* (ISSN 2076-0817) (available at: https://www.mdpi.com/journal/pathogens/special_issues/Leptospira_infections).

For citation purposes, cite each article independently as indicated on the article page online and as indicated below:

LastName, A.A.; LastName, B.B.; LastName, C.C. Article Title. *Journal Name* **Year**, *Article Number*, Page Range.

ISBN 978-3-03943-262-2 (Hbk)
ISBN 978-3-03943-263-9 (PDF)

© 2020 by the authors. Articles in this book are Open Access and distributed under the Creative Commons Attribution (CC BY) license, which allows users to download, copy and build upon published articles, as long as the author and publisher are properly credited, which ensures maximum dissemination and a wider impact of our publications.
The book as a whole is distributed by MDPI under the terms and conditions of the Creative Commons license CC BY-NC-ND.

Contents

To Prof. Domenico Cerri (Head of the Department of Veterinary Science, University of Pisa, Italy) for passing on his legacy and love for microbiology and infectious diseases, especially leptospirosis, to all his colleagues. This Special Issue could have not been published without his precious contribution.

About the Editors

Giovanni Cilia graduated in Veterinary Biotechnological Sciences master's degree in 2016 at the University of Milan (Italy). Since 2017, he has been a Ph.D. student in "Veterinary Sciences" at the Department of Veterinary Sciences of the University of Pisa. He performs support teaching and seminar activities for the courses "Infectious Bacterial Diseases", "Microbiology and Immunology", "Microbiology and Biotechnology Applied to Animal Production", and "Veterinary Hygiene". During his activities, he applied to internships in different research institutes, including the Research Center for Agriculture and Environment (Bologna, Italy) and Biology of Spirochetes Unit, National Reference Center for Leptospirosis of Institute Pasteur (Paris, France). His research is focused on bacterial infectious diseases in domestic animals and wildlife, including leptospirosis, salmonellosis, brucellosis, and E.coli infection and antibiotic-resistances. Moreover, his other research topics include the antibacterial activity of natural substances and infectious diseases in honey bees. He is the author/co-author of several scientific publications in international journals and the Editor of some Infectious Disease Special Issues and Topic Editor of "Veterinary Sciences".

Fabrizio Bertelloni graduated in Veterinary Medicine in 2008 at the University of Pisa (Italy). In 2012, he obtained his specialization in "Animal Health, Breeding and Animal Productions" at the University of Pisa. In 2015, he obtained his Ph.D. in "Veterinary Science" at the University of Pisa. From 2016 to 2018, he undertook a Post-Doc at the Infectious Disease sector of the Department of Veterinary Science, University of Pisa. Since 2018, he has been a Research Fellow (Junior Fixed-Term Research Fellows- RTDA) at the University of Pisa and lecturer in Veterinary Hygiene. His area of research focuses on bacterial infectious disease of wild and domestic animals, in particular salmonellosis, leptospirosis, and brucellosis. He is the author/co-author of more than 40 scientific publications in international journals and a member of the editorial board of " Veterinary Sciences".

Filippo Fratini graduated in Veterinary Medicine in 2002 at the University of Pisa (Italy). In 2005, he obtained a Ph.D. in "Animal production, health and food hygiene in Mediterranean climate countries" and, in 2009, he obtained a specialization in "Animal Health, Breeding and Livestock Production" at the University of Pisa. He first served as a researcher in Infectious Animal Diseases and, since 2017, he has been an Associate Professor at the University of Pisa and lecturer in the courses of "Veterinary Epidemiology" and "Infectious Bacterial Diseases". His main lines of research are focused on infectious diseases caused by bacteria in domestic and wild animals such as leptospirosis, salmonellosis, and paratuberculosis. His other topics of interest include the antibacterial activity of natural substances and their synergistic activities, mastitis, and antibiotic-resistance in its causative agents and microbiological aspects in edible insect rearing. Moreover, he is author/co-author of over 120 scientific publications in national and international journals and a member of the editorial board of "Microbiological Research" and "Evidenced-Based Complementary and Alternative Medicine".

Preface to "*Leptospira* infections in Domestic and Wild Animal"

In the wide range of zoonoses, leptospirosis is one of the most important re-emerging diseases. The increasing interest in leptospirosis all over the world is due to important environmental issues such as climate change and the increasing presence of wild animals in urban areas. The wide range of *Leptospira* hosts and the recent involvement of new species of wild and domestic animals among this group also contribute to the maintenance of *Leptospira* in the environment and promote human infection. Moreover, the growth of the number of extensive and semi-extensive animal farms, is shaping leptospirosis epidemiology, since it favors the environmental spreading of *Leptospira* strains by promoting the contact between wild and domestic animals.

Traditionally, leptospirosis was associated with specific worker categories (such as farmers, veterinarians) or activities such as hunting due to the direct or indirect contact with the urine of a potential *Leptospira* reservoir. However, the increased occasions of contact between animals (wild and domestic) and humans caused a modification in the relation between *Leptospira* serovars and their maintenance-hosts, not only in Europe but also in other parts of the world.

Considering the key role that wildlife and domestic animals play in the maintenance, diffusion, and transmission of leptospirosis, the aim of this book is to collect recent research about the involvement of wild and domestic animals in the epidemiology of leptospirosis.

On this occasion, the Editors would like to warmly thank Prof. Domenico Cerri (Head of the Department of Veterinary Science, University of Pisa, Italy) for passing on his legacy and love for microbiology and infectious diseases, especially leptospirosis, to all his colleagues. This Special Issue could have not been published without his precious contribution.

Giovanni Cilia, Fabrizio Bertelloni, Filippo Fratini
Editors

pathogens

MDPI

Editorial

Leptospira Infections in Domestic and Wild Animals

Giovanni Cilia, Fabrizio Bertelloni * and Filippo Fratini

Department of Veterinary Sciences, University of Pisa, Viale delle Piagge 2, 56124 Pisa, Italy; giovanni.cilia@vet.unipi.it (G.C.); filippo.fratini@unipi.it (F.F.)
* Correspondence: fabrizio.bertelloni@unipi.it; Tel.: +39-050-2216-969

Received: 9 July 2020; Accepted: 14 July 2020; Published: 15 July 2020

check for updates

Abstract: Leptospirosis is a worldwide-distributed, re-emerging zoonosis due to the large variety of wild and domestic animal species that can play the role of natural or accidental host. Currently, specific animal species play an important role as the reservoir for particular *Leptospira* serovars, although recent investigations have highlighted new host–pathogen interactions involved in *Leptospira* epidemiology. Furthermore, the constant modification of ecosystems and wildlife habitats and the constantly increasing number of animal species moving towards urban or peri-urban areas are increasing the possibility of direct or indirect contacts between wildlife and domestic animals; furthermore, the constant modification of animal leptospirosis also causes problems for human health. The studies published in this Special Issue have evidenced and confirmed the hidden role of a large variety of animal species, domestic and wild, in the leptospirosis epidemiology. They highlighted the necessity for continuous monitoring and large-scale surveillance studies to better understand this neglected and re-emerging zoonosis.

Keywords: *Leptospira*; leptospirosis; pathology; zoonosis; wildlife; infectious disease; wild boar; *Sus* scrofa; African green monkey; *Chlorocebus sabeus*; vaccine; dogs; bovine; new Pomona serovars; MLST; VNTR

Introduction

Leptospira spp. is a Gram-negative bacterium belonging to Spirochetales order. *Leptospira* species, with more than 260 antigenically distinct serovars, have been grouped into pathogenic, intermediate, and saprophytic, with different levels of pathogenicity for animals and humans [1]. Pathogenic species of *Leptospira* are causes of leptospirosis, a re-emerging and widespread zoonosis. Moreover, intermediate *Leptospira* may be potentially pathogens and are responsible for mild infection, while saprophytic ones are spread in the environment and they are nonpathogenic [2,3].

Leptospirosis occurs in tropical, subtropical, and temperate zones, representing a public health problem due to its involving of humans, domestic and wild animals, which can be maintenance or accidental hosts [4,5]. Maintenance hosts are asymptomatic renal carriers, that contribute to maintaining and sharing the infection-shedding *Leptospira* with urine in the environment. Accidental contact with *Leptospira*-infected urine represents the first cause of infection, that could produce clinical disease [1]. So, the *Leptospira* epidemiology is closely related to the maintenance host species, generally linked to a specific *Leptospira* serovar; usually Icterohaemorrhagiae and Ballum serogroups are associated with rodents [6–9], Pomona and Tarassovi serogroups with pigs and wild boar [10–13], Bratislava serogroup with horses [14,15] and Sejroe serogroup with bovines and ovine [16,17].

To better understand the epidemiology of the *Leptospira* infection in wild and domestic animals, this Special Issue aims to bring together research studies related to investigating the role of these animals in leptospirosis epidemiology, as well as new prospective for treatment and prevention. The seven studies published in this Special Issue highlighted a wide-spectrum of *Leptospira* hosts in relation to different geographic areas and they emphasized new host–pathogen interactions in common and uncommon animal species.

Bertasio et al. performed a *Leptospira* investigation on 131,660 pigs sampled from 2002 to 2017 from 4715 farms in Northern Italy. A serological positivity rate of 13.05% was determined through Microscopic Agglutination Test (MAT). Australis was the most frequently identified serogroup (77.29%), followed by Pomona (18.47%), Tarassovi (1.51%) and Icterohaemorrhagiae (1.40%). Moreover, culture isolation and RealTime PCR were performed on 347 kidneys and 470 clinical samples, respectively. Using a Multi-Locus sequence typing (MLST) and Variable-Number Tandem Repeat (VNTR) analysis, 43 samples produced identical profiles but, after 2014, three new *Leptospira interrogans* serogroup Pomona genotypes were observed. Moreover, two isolates showed new MLST profiles and an unclassified identification by monoclonal antibodies [18].

Balboni and colleagues characterized *Leptospira* isolated in dogs with confirmed symptomatic acute leptospirosis. *Leptospira* spp. DNA were detected in urine, blood, or both in samples from nine infected dogs; obtained isolates were analyzed using the MLST technique. The isolates from two dogs were successfully typed: one belonging to Sequence Type (ST) 17 and one to ST198, *Leptospira interrogans* serogroups Icterohaemorrhagiae and Australis, respectively. The study provided the first molecular analysis aimed at identifying infectious *Leptospira* directly on DNA from biological samples of dogs. The authors showed that serogroup Australis could lead to a severe clinical presentation of leptospirosis in dogs [19].

Cilia et al. carried out an investigation aimed at evaluating the prevalence of *Leptospira* with serological, bacteriological, and molecular assays in wild boar (*Sus scrofa*) hunted in Tuscany (Italy) during two hunting seasons. In total, 287 specimens of sera, kidneys, and liver were collected to perform MAT, isolation, and RealTime PCR to detect pathogenic (*lipL32* gene), intermediate (*16S rRNA* gene), and saprophytic (*23S rRNA* gene) *Leptospira*. Within the sera, 39 (13.59%) were positive to MAT, and Australis was the most represented serogroup (4.88%), followed by Pomona (4.18%), and Tarassovi (3.14%). Moreover, four *Leptospira* cultures were positive and isolates were subjected to MLST analysis; one of them was identified as *L. borgpetersenii* serovar Tarassovi, and three as *L. interrogans* serovar Bratislava. Pathogenic *Leptospira* DNA were detected in 32 wild boar kidneys (11.15%). The characterization through the amplification of the *rrs2* gene highlighted they were positive for *L. interrogans* (23 kidneys), *L. borgpetersenii* (4 kidneys), and *L. kirschneri* (1 kidney), while nine kidneys (3.14%) were positive for intermediate *Leptospira*, all belonging to *L. fainei* [20].

Gregoire et al. investigated bovine leptospirosis. During July 2014, an emerging phenomenon of increased incidence of icteric abortions associated with leptospiral infection occurred in southern Belgium. First-line serological analyses targeting cattle-adapted serovars failed at initial diagnosis. This study provided a comprehensive description of laboratory findings—at the level of necropsy, serology and molecular diagnosis—regarding icteric and non-icteric abortions (n = 116) recorded during this time (years 2014–2015) and associated with incidental infection by serovars such as Grippotyphosa, Australis and Icterohaemorrhagiae. Based on these tests, a diagnostic pathway is proposed for these types of infection in cattle to establish an affordable but accurate diagnosis in the future. The investigations add insight into the understanding of the pathogenesis of bovine leptospirosis associated with serovars classically described as no maintenance host [21].

Rajeev and colleagues investigated the potential asymptomatic *Leptospira* carrier status among African green monkeys (*Chlorocebus sabeus*) on the Caribbean island of Saint Kitts. Moreover, the authors analyzed if any renal pathology could be associated with *Leptospira* exposure. Of the African green monkeys tested, 48% were positive for *Leptospira* antibodies by the microscopic agglutination test. *Leptospira* DNA was detected in 4% of kidney samples tested, using a *lipl32* gene-based PCR. Microscopic renal lesions, from minimal to severe, were evidenced in 85% of the African green monkey kidneys. Most of the African green monkeys (n = 26) had only minimal to mild interstitial nephritis and a few (n = 3) had moderate to severe lesions. The presence of interstitial nephritis was not significantly associated with *Leptospira* exposure [22].

Bertasio and colleagues detected and genotyped *Leptospira* isolated from symptomatic dogs in northeast Italy between 2013 and 2019. In total, 1631 dogs were tested using RealTime PCR, and the isolates from 193 dogs were subjected to MLST and a VNTR analysis. The bacteria were successfully isolated from 15 symptomatic dogs. Six distinct STs were found for 135 isolates, with 3 STs characterizing *Leptospira interrogans* (ST17, ST198 and ST24), 2 STs characterizing *Leptospira kirschneri* (ST117 and ST289) and 1 ST characterizing *Leptospira borgpetersenii* (ST155), revealing the circulation of the serogroups Icterohaemorrhagiae, Australis, Sejroe and Pomona. The genomic analysis of 17 samples did not result in any additional discrimination. Genotypes were compared with those of strains present in the historical internal database, and possible transmission chains were identified from rat, mouse, hedgehog, and pig. The work highlights the importance of molecular methods in revealing and identifying circulating *Leptospira* strains, and it also encourages the evaluation of the ability of commercially available vaccines to reduce the disease burden among dogs [23].

Zakharova et al. assessed the potential risk expositing to the infection as a function of environmental determinants in the Republic of Sakha (Yakutia), Russian Federation. The authors applied environmental niche modeling using leptospirosis cases in livestock and wild animals in the period from 1995 to 2019 with regard to a set of landscape, climatic, and socioeconomic variables, both for the current climate and for the projected climate for 2041–2060. The MaxEnt model performed well (AUC = 0.930), with the mean temperature of the warmest quarter, mean diurnal range, land cover type, and altitude being the most contributing variables. Consequent zoning based on the proportion of high-risk cells within each administrative unit suggested that five out of the 36 districts of the Republic are at high risk in the current climate conditions, with three more districts expected to demonstrate a high risk by 2060. The study presents the first ever attempt at leptospirosis ecological modeling in Russia. Its results correspond well to the findings of other authors and underline the importance of considering ecological factors when conducting a leptospirosis risk assessment [24].

Author Contributions: G.C., F.B. and F.F. conceptualized, wrote, and reviewed the manuscript. All authors have read and agreed to the published version of the manuscript.

Funding: This research received no external funding.

Acknowledgments: Authors are grateful to all authors and reviewers that participated in this Special Issue.

Conflicts of Interest: The authors declare no conflict of interest.

References

1. Adler, B.; de la Peña Moctezuma, A. Leptospira and leptospirosis. *Vet. Microbiol.* **2010**, *140*, 287–296. [CrossRef]
2. Vincent, A.T.; Schiettekatte, O.; Goarant, C.; Neela, V.K.; Bernet, E.; Thibeaux, R.; Ismail, N.; Mohd Khalid, M.K.N.; Amran, F.; Masuzawa, T.; et al. Revisiting the taxonomy and evolution of pathogenicity of the genus Leptospira through the prism of genomics. *PLoS Negl. Trop. Dis.* **2019**, *13*, e0007270. [CrossRef]
3. Guglielmini, J.; Bourhy, P.; Schiettekatte, O.; Zinini, F.; Brisse, S.; Picardeau, M. Genus-wide Leptospira core genome multilocus sequence typing for strain taxonomy and global surveillance. *PLoS Negl. Trop. Dis.* **2019**, *13*, e0007374. [CrossRef]
4. Bertelloni, F.; Cilia, G.; Turchi, B.; Pinzauti, P.; Cerri, D.; Fratini, F. Epidemiology of leptospirosis in North-Central Italy: Fifteen years of serological data (2002–2016). *Comp. Immunol. Microbiol. Infect. Dis.* **2019**, *65*, 14–22. [CrossRef]
5. Picardeau, M. Virulence of the zoonotic agent of leptospirosis: Still terra incognita? *Nat. Rev. Microbiol.* **2017**, *15*, 297–307. [CrossRef] [PubMed]
6. Blasdell, K.R.; Morand, S.; Perera, D.; Firth, C. Association of rodent-borne Leptospira spp. with urban environments in Malaysian Borneo. *PLoS Negl. Trop. Dis.* **2019**, *13*, e0007141. [CrossRef] [PubMed]
7. Mori, M.; Bourhy, P.; Le Guyader, M.; Van Esbroeck, M.; Djelouadji, Z.; Septfons, A.; Kodjo, A.; Picardeau, M. Pet rodents as possible risk for leptospirosis, Belgium and France, 2009 to 2016. *Eurosurveillance* **2017**, *22*. [CrossRef] [PubMed]

8. Coppola, F.; Cilia, G.; Bertelloni, F.; Casini, L.; D'Addio, E.; Fratini, F.; Cerri, D.; Felicioli, A. Crested porcupine (Hystrix cristata L.): A new potential host for pathogenic Leptospira among semi-fossorial mammals. *Comp. Immunol. Microbiol. Infect. Dis.* **2020**, *70*, 101472. [CrossRef] [PubMed]
9. Cilia, G.; Bertelloni, F.; Coppola, F.; Turchi, B.; Biliotti, C.; Poli, A.; Parisi, F.; Felicioli, A.; Cerri, D.; Fratini, F. Isolation of *Leptospira* serovar Pomona from a crested porcupine (*Hystrix cristata*, L., 1758). *Vet. Med. Sci.* **2020**. [CrossRef] [PubMed]
10. Bertelloni, F.; Mazzei, M.; Cilia, G.; Forzan, M.; Felicioli, A.; Sagona, S.; Bandecchi, P.; Turchi, B.; Cerri, D.; Fratini, F. Serological Survey on Bacterial and Viral Pathogens in Wild Boars Hunted in Tuscany. *Ecohealth* **2020**, *17*, 85–93. [CrossRef] [PubMed]
11. Cilia, G.; Bertelloni, F.; Mignone, W.; Spina, S.; Berio, E.; Razzuoli, E.; Vencia, W.; Franco, V.; Cecchi, F.; Bogi, S.; et al. Molecular detection of Leptospira spp. in wild boar (Sus scrofa) hunted in Liguria region (Italy). *Comp. Immunol. Microbiol. Infect. Dis.* **2020**, *68*, 101410. [CrossRef]
12. Vale-Goncalves, H.M.; Cabral, J.A.; Faria, M.C.; Nunes-Pereira, M.; Faria, A.S.; Veloso, O.; Vieira, M.L.; Paiva-Cardoso, M.N. Prevalence of *Leptospira* antibodies in wild boars (*Sus scrofa*) from Northern Portugal: Risk factor analysis. *Epidemiol. Infect.* **2015**, *143*, 2126–2130. [CrossRef] [PubMed]
13. Boqvist, S.; Bergström, K.; Magnusson, U. Prevalence of Antibody to Six Leptospira Servovars in Swedish Wild Boars. *J. Wildl. Dis.* **2012**, *48*, 492–496. [CrossRef] [PubMed]
14. Arent, Z.; Frizzell, C.; Gilmore, C.; Allen, A.; Ellis, W.A. Leptospira interrogans serovars Bratislava and Muenchen animal infections: Implications for epidemiology and control. *Vet. Microbiol.* **2016**, *190*, 19–26. [CrossRef]
15. Rocha, T.; Ellis, W.A.; Montgomery, J.; Gilmore, C.; Regalla, J.; Brem, S. Microbiological and serological study of leptospirosis in horses at slaughter: First isolations. *Res. Vet. Sci.* **2004**, *76*, 199–202. [CrossRef] [PubMed]
16. Farina, R.; Cerri, D.; Renzoni, G.; Andreani, E.; Mani, P.; Ebani, V.; Pedrini, A.; Nuvoloni, R. Leptospira interrogans in the genital tract of sheep. Research on ewes and rams experimentally infected with serovar hardjo (hardjobovis). *New Microbiol.* **1996**, *19*, 235–242. [PubMed]
17. Cerri, D.; Ebani, V.V.; Fratini, F.; Pinzauti, P.; Andreani, E. Epidemiology of leptospirosis: Observations on serological data obtained by a "diagnostic laboratory for leptospirosis" from 1995 to 2001. *New Microbiol.* **2003**, *26*, 383–389.
18. Bertasio, C.; Papetti, A.; Scaltriti, E.; Tagliabue, S.; D'Incau, M.; Boniotti, M.B. Serological Survey and Molecular Typing Reveal New Leptospira Serogroup Pomona Strains among Pigs of Northern Italy. *Pathogens* **2020**, *9*, 332. [CrossRef]
19. Balboni, A.; Zamagni, S.; Bertasio, C.; Boniotti, M.B.; Troìa, R.; Battilani, M.; Dondi, F. Identification of Serogroups Australis and Icterohaemorrhagiae in Two Dogs with a Severe Form of Acute Leptospirosis in Italy. *Pathogens* **2020**, *9*, 351. [CrossRef]
20. Cilia, G.; Bertelloni, F.; Angelini, M.; Cerri, D.; Fratini, F. Leptospira Survey in Wild Boar (Sus scrofa) Hunted in Tuscany, Central Italy. *Pathogens* **2020**, *9*, 377. [CrossRef]
21. Grégoire, F.; Bakinahe, R.; Petitjean, T.; Boarbi, S.; Delooz, L.; Fretin, D.; Saulmont, M.; Mori, M. Laboratory Diagnosis of Bovine Abortions Caused by Non-Maintenance Pathogenic Leptospira spp.: Necropsy, Serology and Molecular Study Out of a Belgian Experience. *Pathogens* **2020**, *9*, 413. [CrossRef] [PubMed]
22. Rajeev, S.; Bolfa, P.; Shiokawa, K.; Beierschmitt, A.; Palmour, R. Leptospira Infection in African Green Monkeys in an Endemic Area: An Opportunity for Comparative Studies in a Natural Environment. *Pathogens* **2020**, *9*, 474. [CrossRef] [PubMed]
23. Bertasio, C.; Boniotti, M.B.; Lucchese, L.; Ceglie, L.; Bellinati, L.; Mazzucato, M.; Furlanello, T.; D'Incau, M.; Natale, A. Detection of New Leptospira Genotypes Infecting Symptomatic Dogs: Is a New Vaccine Formulation Needed? *Pathogens* **2020**, *9*, 484. [CrossRef] [PubMed]
24. Zakharova, O.I.; Korennoy, F.I.; Toropova, N.N.; Burova, O.A.; Blokhin, A.A. Environmental Risk of Leptospirosis in Animals: The Case of the Republic of Sakha (Yakutia), Russian Federation. *Pathogens* **2020**, *9*, 504. [CrossRef]

© 2020 by the authors. Licensee MDPI, Basel, Switzerland. This article is an open access article distributed under the terms and conditions of the Creative Commons Attribution (CC BY) license (http://creativecommons.org/licenses/by/4.0/).

pathogens

MDPI

Article

Leptospira Infection in African Green Monkeys in an Endemic Area: An Opportunity for Comparative Studies in a Natural Environment

Sreekumari Rajeev [1,*,†], Pompei Bolfa [1], Kanae Shiokawa [1,‡], Amy Beierschmitt [1,2] and Roberta Palmour [2,3]

1 School of Veterinary Medicine, Ross University, Basseterre KN 0101, Saint Kitts and Nevis; pbolfa@rossvet.edu.kn (P.B.); Kanae.Shiokawa@bio-techne.com (K.S.); ABeierschmitt@rossvet.edu.kn (A.B.)
2 Behavioral Science Foundation, Basseterre KN 0101, Saint Kitts and Nevis; Roberta.palmour@mcgill.ca
3 Departments of Human Genetics and Psychiatry, Faculty of Medicine, McGill University, Montréal, QC H3A 1A1, Canada
* Correspondence: s.rajeev@ufl.edu; Tel.: +1-352-294-4158
† Present address: College of Veterinary Medicine, University of Florida, Gainesville, FL 32610, USA.
‡ Present address: Bio-Techne, Minneapolis, MN 55413, USA.

Received: 8 May 2020; Accepted: 11 June 2020; Published: 16 June 2020

Abstract: This study was performed to investigate the potential asymptomatic *Leptospira* reservoir status among African green monkeys (AGMs) in the Caribbean island of Saint Kitts, and whether there is any renal pathology associated with *Leptospira* exposure. Forty-eight percent of AGMs tested were positive for *Leptospira* antibodies by the microscopic agglutination test. *Leptospira* DNA was detected in 4% of kidney samples tested using a *lipl*32 gene based PCR. We observed minimal to severe microscopic renal lesions in 85% of the AGM kidneys evaluated. The majority of the AGMs (n = 26) had only minimal to mild interstitial nephritis and a few (n = 3) had moderate to severe lesions. The presence of interstitial nephritis was not significantly associated with *Leptospira* exposure. The presence of infected AGMs in a small surface limited geographic region may pose zoonotic threat to humans and animals. The impact of *Leptospira* infection in renal pathology in AGMs warrants further investigation. AGMs residing in a natural setting in an insular, surface limited *Leptospira* endemic geographic region may offer opportunities for comparative studies to advance the field of leptospirosis. Due to their similarity to humans, such studies in AGMs may also provide translational opportunities to advance *Leptospira* research.

Keywords: *Leptospira*; African green monkeys; Caribbean; renal lesions; zoonosis

1. Introduction

Leptospirosis is a widespread zoonotic disease of global importance and may cause life-threatening illness in humans and animals. It is a prime example of the "one health" paradigm, as the interface between humans, animals, and the environment plays a major role in the maintenance and transmission of the causative bacteria *Leptospira*. Leptospirosis can be caused by numerous pathogenic serovars of spirochete bacteria, *Leptospira*, that are maintained in the renal tubules of animals and the environment [1,2]. In animals, in addition to clinical disease, *Leptospira* infection results in asymptomatic renal colonization leading to urinary shedding and subsequent contamination of the environment allowing the transmission to susceptible hosts. Various reports confirm the susceptibility of nonhuman primates to leptospirosis. Fulminant leptospirosis was reported in a captive squirrel monkey colony in French Guiana presenting with acute illness, jaundice, hemorrhagic syndrome, and mortality [3]. An epizootic event was reported in a number of captive capuchin monkeys housed in a Columbia wildlife rehabilitation center, suggesting the potential for new reservoirs and transmission routes for *Leptospira*, threatening conservation efforts and public health [4]. In the outbreak reported, jaundice and pulmonary hemorrhage were the major manifestations resulting in 71% morbidity and 27% mortality among infected animals. *Leptospira* seroprevalence has been reported in other nonhuman primates in various geographic regions [5,6]. In experimental infections, severe clinical signs similar to those seen in humans have been reported in nonhuman primates. Experimental infection in grivet monkeys has resulted in severe lymphocytic meningoencephalitis and inflammation of kidneys [7,8]. Patterns comparable to those seen in the severe forms of human leptospirosis, including pulmonary hemorrhage and severe tubulointerstitial nephritis, were present in experimental infections in marmoset monkeys [9]. Recently, we have documented *Leptospira* seroprevalence in captive and wild African green monkey (*Chlorocebus sabeus)* population inhabiting the Caribbean island of Saint Kitts, suggesting exposure to various *Leptospira* serovars [10]. In an in-depth systematic review published in 2015, a high estimated *Leptospira* related human morbidity (50.68) and mortality (2.9) per 100,000 population was reported in the Caribbean region, ranking third among WHO GBD(Global Burden of Disease) regions [11]. In a country level estimate, Saint Kitts has an estimated *Leptospira*-related human morbidity and mortality rate of 132. 71 and 7, respectively [11]. Proximity to the *Leptospira* infected animals can be a potential risk factor for other animals and humans. Our studies have also documented *Leptospira* infection and exposure in other animal species on the island [12–16]. In this pilot study, we investigated whether the African green monkeys (AGMs) could potentially serve as an asymptomatic source of infection and whether there is any association between *Leptospira* infection or exposure and the presence of renal pathology in a set of AGMs.

2. Results

This study was conducted at Saint Kitts, a small Caribbean island located in the Lesser Antilles. The AGMs (n = 83) sampled included 49 adults, 8 young adults, and 10 juveniles. Out of these, 50 were female, 23 were male, and sex of 10 was unavailable. All the AGMs were apparently healthy at the time of sample collection. We tested a total of 83 AGM serum samples by microscopic agglutination test (MAT). From this cohort, 50 AGM kidney samples were available for PCR and 34 were available for histopathology examination. A list of serovars used in the MAT panel and the results are shown in the Table 1. All the samples with 50% agglutination were recorded as positive for the presence of antibodies in the screening assay at a final dilution of 1:50 and MAT titers were determined for all the positive samples.

Table 1. *Leptospira* species, serogroups and serovars of strains used in the microscopic agglutination test (MAT) panel. The MAT results for each of the serovar are given in parenthesis.

Species	Serogroup	Serovar (Results)
L. interrogans	Australis	Australis (Negative) Bratislava (Positive)
	Autumnalis	Autumnalis (Positive)
	Bataviae	Bataviae (Positive)
	Canicola	Canicola (Negative)
	Djasiman	Djasiman (Positive)
	Grippotyphosa	Grippotyphosa (Negative)
	Icterohemorrhagiae	Icterohemorrhagiae (Positive) Mankarso (Positive) Copenhageni (Positive)
	Pomona	Pomona (Positive)
	Sejroe	Hardjo (Negative) Wolffii (Negative)
L. borgpetersenii	Ballum	Ballum (Positive)
	Javanica	Javanica (Negative)
	Tarrasovi	Tarrasovi (Negative)
L. santarosai	Pyrogenes	Alexi (Negative) Pyrogenes (Negative)
	Mini	Georgia (positive)
L. kirschneri	Cynopteri	Cynopteri (Positive)
L. weilii	Celledoni	Celledoni (Positive)

Thirty nine out of 83 (47%) serum samples tested by MAT were positive for aggluntinating antibodies to one or more *Leptospira* serovars. MAT test was positive for 12 serovars from the MAT panel and the MAT titers ranged from 50–1600. Predominant serovars included, Bataviae (15/83:18%), Mankarso (13/83:16%), Ballum (12/83:15%), and Icterohaemorrhagiae (10/83:12%). Number of samples positive for each of the tested *Leptospira* serovars in the MAT panel is shown in Figure 1.

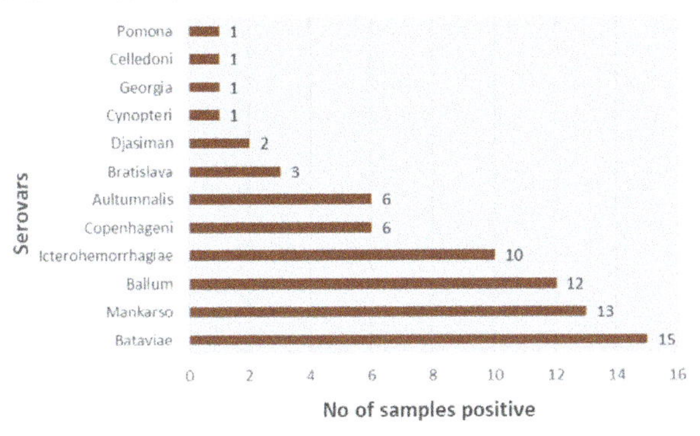

Figure 1. Summary of MAT results showing number of samples positive for *Leptospira* serovars.

Agglutinating antibodies to multiple serovars were observed in the same AGMs. The majority of the samples ($n = 21$)were positive for only one serovar. Few samples had MAT titers to two serovars ($n = 9$), three serovars ($n = 6$), four serovars ($n = 2$), and six serovars ($n = 1$). Table 2 lists the MAT results from 39 individual AGM samples showing positive serovars and corresponding titers.

Table 2. Distribution of positive serovars and their corresponding MAT titers (in parenthesis) in 39 individual African green monkeys (AGMs) tested positive by MAT. AGM# in the column 1 represents the individual identification number of the AGMs tested.

AGM #	Serovars Positive (MAT-Titer)
3	Autumnalis (200), Bratislava (1600)
4	Ballum (200)
5	Bataviae (100)
7	Ballum (100)
10	Ballum (400)
14	Ballum (50)
15	Ballum (50)
32	Djasiman(50)
36	Icterohemorrhagiae (200), Mankarso (50)
37	Icterohemorrhagiae (200), Mankarso(100), Copenhageni (100)
38	Bataviae (100)
39	Bataviae (100)
40	Icterohemorrhagiae (100) Mankarso(50)
41	Icterohemorrhagiae (50)
42	Bataviae (100)
43	Icterohemorrhagiae (200), Mankarso (100), Ballum (400), Bataviae (100)
45	Icterohemorrhagiae (50), Bataviae (100)
46	Icterohemorrhagiae (400), Mankarso (400), Copenhageni (200)
47	Bataviae (100)
48	Bataviae (200)
51	Bataviae (50)
53	Icterohemorrhagiae (50), Mankarso(50)
55	Bataviae (100)
56	Ballum (100)
57	Icterohemorrhagiae (400), Mankarso (800), Copenhageni (100)
66	Autumnalis (50)
67	Autumnalis (50)
69	Autumnalis (200), Cynopteri (50), Georgia (50), Icterohemorrhagiae (100), Mankarso (200), Copenhageni (100)

Table 2. *Cont.*

AGM #	Serovars Positive (MAT-Titer)
70	Bratislava (50)
71	Mankarso (100), Copenhageni (50)
72	Autumnalis (50), Mankarso (100)
73	Ballum (200), Bataviae (100),Celledoni (50), Mankarso (100)
74	Autumnalis (400), Bratislava (50), Pomona (50)
76	Ballum (100), Bataviae (100), Mankarso (50)
77	Bataviae (50)
78	Ballum (100)
79	Ballum (200), Bataviae (100)
81	Ballum (50) Mankarso (50), Copenhageni (50)
83	Bataviae (50), Djasiman (50)

Leptospira DNA was detected in 4% (2/50) kidney samples processed for PCR, targeting the lipL32 gene with a cycle threshold value of 34.16 and 36.83. One (AGM #69) of these PCR positive samples was positive by MAT. Out of these 50 kidney samples, we examined 34 formalin-fixed and hematoxylin and eosin (H&E) stained kidney samples by routine histopathology examination. Minimal to severe lymphoplasmacytic interstitial nephritis was seen in 29/34 (85%) of the AGM kidneys. The lesions were multifocal in all affected animals, mainly seen in the cortex, with various degrees of intensity: minimal (n = 12), mild (n = 14), moderate (n = 2), and severe (n = 1). In the moderate and severe cases, various degrees of tubular degeneration, loss, and regeneration were seen. In the mild cases, focal interstitial fibrosis was seen in one sample, and multifocal tubulointerstitial mineralization (von Kossa positive- confirming the presence of Ca) was seen in nine cases. In one AGM, calcium oxalate crystals were seen in the tubular epithelial cells and tubular lumens. In the kidneys from one monkey, there were radiating arrays of pale eosinophilic, acellular spicules surrounded by macrophages and multinucleated giant cells highly suggestive of medullary tophi. There was no significant association (p = 1) between positve MAT results and the presence of interstitial nephritis (Table 3). Figure 2 illustrates representative lesions observed in the kidneys.

Table 3. Comparison of MAT results with the presence of interstitial nephritis using Fisher's exact test. MAT—microscopic agglutination test; ISN—interstitial nephritis. A p-value less than 0.05 is considered as significant.

	ISN+	ISN−	Total
MAT+	7	1	8
MAT−	22	4	26
Total	29	5	34

(p = 1).

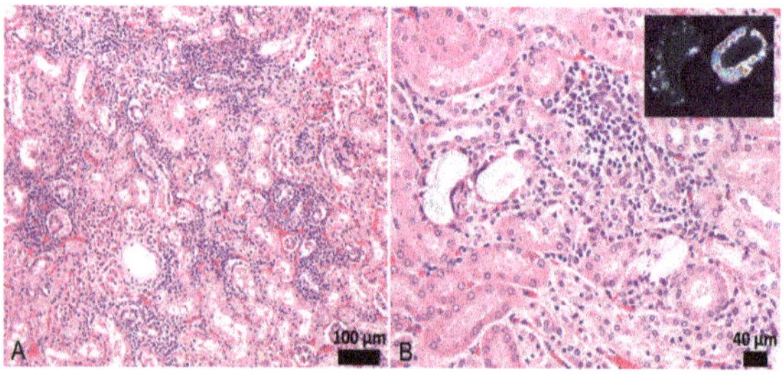

Figure 2. (**A**,**B**) Renal lesions in AGMs. (**A**) Multifocal lymphoplasmacytic interstitial nephritis and mild tubular epithelial degeneration. (**B**) Focal lymphoplasmacytic interstitial nephritis and the presence of intratubular calcium oxalate crystals with loss of epithelial cells and foreign body reaction with multinucleated giant cells. Inset shows crystal birefringence on polarized light.

3. Discussion

African green monkeys, also known as vervet monkeys, are a highly adaptable nonhuman primate species inhabiting sub-Saharan Africa. AGMs were introduced in the Caribbean region by the European settlers in the 17th century and are widespread in the West Indian islands of Barbados, Saint Kitts, Nevis, and Saint Martin. In Saint Kitts, the wild population has increased drastically over the years since its first introduction [17]. The captive animals are mainly used for entertaining tourists and biomedical research. The free-roaming AGM population observed in clusters in human-inhabited areas of the island have destructive habits and a negative impact on the island's agriculture. Trapping free-roaming AGMs is a common practice among local people, and some of these wild-caught animals are used as a source of bush meat. In Saint Kitts, a *Leptospira* endemic area, the infected AGMs, and the environment contaminated with their urine may pose a significant risk factor for humans and animals in contact.

In this study, *Leptospira* renal infection was confirmed in only two of the kidney samples tested by PCR. Only one of these samples was evaluated by histopathology, but only mild multifocal interstitial nephritis was observed in this case and this animal was negative by microscopic agglutination test. This is not surprising as we have observed similar findings in other animal species where antibody response or lesions were minimal despite the presence of organisms [12,13,18]. It is roughly estimated that 50,000 AGMs inhabit Saint Kitts, a small island with a 168 km^2 area. Although a low-level asymptomatic infection was observed in the AGM population tested, the presence of large AGM population density in a surface area-limited geographic location is concerning. At a given time point, the magnitude of contamination and the potential risk of transmission can be proportionally high, even when a low number of animals are shedding. The handling of contaminated crops has resulted in recent Leptospirosis outbreaks in agriculture workers in Europe and Australia [19,20]. A similar zoonotic threat may exist in this island, warranting biosecurity measures in controlling contamination of agriculture products with AGM urine.

Exposure to *Leptospira,* as evidenced by the presence of agglutinating antibodies, in this cohort of AGMs is in agreement with our previous study [10]. The highest exposure was documented to serovar Bataviae, and the highest MAT titer was observed for serovar Bratislava. High-level exposure to these serovars was also observed in our previous study. Exposure to members of the serogroup Icterohemorrhagiae (serovars Icterohemorrhagiae, Mankarso, Copenhageni) was found in the population tested. We presume this to be most likely due to the cross-reactivity between members of the same serogroup. Out of two PCR-positive samples one was positive for six *Leptospira* serovars with MAT titers ranging from 50–200, suggestive of a potential early active infection. We have isolated *L. interrogans* serovar Copenhageni from rats, mongooses, and a dog from the island [12,13,15]. In rats from which we isolated *L. interrogans* serovar Copenhageni, the highest exposure was documented for serovar Icterohemorrhagiae but not for the infecting serovar Copenhageni. Exposure to serovar Ballum, a commonly found serovar in Saint Kitts rats [13] and mice (unpublished data), was also observed in this group of AGMs tested. The wild and captive AGMs are in contact with the rodents, and direct or indirect transmission from these highly infected vectors is possible in AGMs. The predominant serovar Bataviae has not been isolated yet from the island. The determination of infecting *Leptospira* serogroups cannot be extrapolated from MAT response alone because of cross-reactivity and sometimes even the absence of antibody response to infecting serovars occurring in some animal species. Cross-reactivity can be observed occasionally in some clinical cases even to unrelated serovars, a phenomenon described as paradoxical reactions [21]. The aspects mentioned above challenges the interpretation of MAT results.

Regular culturing and documentation of common *Leptospira* serovars in a specific geographic location is ideal, but unfortunately, cultures are not attempted in many prevalence studies due to the time consuming, laborious, and often futile nature of this procedure. We were not successful in isolating *Leptospira* from AGM kidney samples.

Mild to severe renal lesions were present in a proportion of AGMs tested. However, we did not find any significant association between *Leptospira* exposure and interstitial nephritis, a typical renal lesion observed in *Leptospira* infection. Recently, widespread renal lesions were also reported in another invasive species, small Indian mongooses inhabiting the island. However, a low level antibody prevalence to *Leptospira* and absence of antibody response, despite colonization in some animals, were observed in this species [18]. *Leptospira* colonizes the renal tubules and may induce host inflammatory response that may favor the removal of the pathogen. The interstitial nephritis might be an effect of previous unsuccessful colonization or might also be the response to other infectious agents. In animals, immune response to infection may have dichotomous outcomes at the humoral or cellular level, and this response may be influenced at the level of host species, individual animals, and the environment. Further detailed investigations will be needed to unravel the mechanisms leading to a likely divergent antibody vs. cell-mediated immune response and the cause of the renal lesions in AGMs.

Acute leptospirosis presented with severe kidney injury, if not diagnosed and treated, may predispose infected hosts to chronic kidney disease that may gradually proceed to end-stage kidney disease. In asymptomatic infection, bacteria may initiate inflammation that might insidiously progress to chronicity. Common *Leptospira* related renal lesions include interstitial nephritis, tubular necrosis, and degeneration that might gradually turn to chronic lesions, such as interstitial fibrosis and tubular atrophy. Chronic kidney disease of unknown origin has been reported as an endemic nephropathy across the globe [22,23]. Asymptomatic *Leptospira* colonization in renal tissue is an overlooked cause of this condition. *Leptospira* is implicated as a cause of chronic renal conditions in humans in some geographic regions and as a risk factor for nephropathy in Sri Lanka and the Mesoamerican region [22,23]. It is reasonable to hypothesize that in endemic areas, in a susceptible host, *Leptospira* infection and subsequent renal colonization can initiate an inflammatory response in the kidneys that develop to chronicity and gradual renal impairment. Therefore, *Leptospira* infection, as an emerging predisposing etiology of chronic kidney disease, warrants in-depth investigation. Chronic lymphoplasmacytic interstitial nephritis, which is not uncommon in kidneys of animals, can have many possible causes, and could be primarily inflammatory or represent foci of antigen persistence. Although the histological lesions were not deemed severe enough in most of our cases to result in clinical symptoms, mild and chronic nature suggests an attack from an infectious agent and subsequent attempt to remove the agent. The renal mineralization, presence of oxalate crystals, and tophi observed in some AGM kidneys can be considered as incidental background lesions, not directly linked to *Leptospira* infection.

AGMs are widely used for biomedical research to study human diseases [24]. They share a high degree of homology with humans and hence may serve as a useful translational model for studying leptospirosis and other diseases in a natural setting. A surface-limited insular location such as Saint Kitts, where *Leptospira* infection is highly endemic in animal populations, offers an ideal opportunity for studying various aspects of leptospirosis. We propose to undertake well-designed prospective longitudinal studies to advance our understanding of the impact of *Leptospira* infection in renal pathology in AGMs. These studies may provide translational knowledge to support *Leptospira*-related chronic kidney disease in the human population. Finally, research studies in a susceptible population in its natural environment might be useful to acquire a deeper understanding of multiple aspects of *Leptospira* infection dynamics, including maintenance, transmission, and pathogenesis. Such studies will facilitate an additional opportunity to elucidate whether *Leptospira* infection has a potential role in conditions such as chronic nephropathy reported in humans in various parts of the globe. This knowledge will ultimately benefit in the development of better prevention and control strategies for *Leptospira* infection and potential health-related consequences in human and animal populations.

4. Materials and Methods

Since this study was initiated on a pilot basis, we used a convenience sampling strategy based on the availability of samples for testing. All the AGM samples originated from Behavioral Science Foundation, Saint Kitts. These include captive animals (n = 49) residing in open cages at the facility and wild animals ($n = 34$) captured from the island and housed temporarily in the facility. All the animal procedures were performed with high ethical standards adhering to IACUC approved protocol (#TSU1.23.18) of the Ross University School of Veterinary Medicine (RUSVM).

Serum samples were obtained from 83 AGMs, and kidney samples were collected from 50 euthanized AGMs. The serum samples were screened against 21 *Leptospira* serovars (shown in Table 1) belonging to 16 serogroups using the microscopic agglutination test (MAT). The MAT was conducted using protocols described previously with modifications [10,25]. Briefly, *Leptospira* serovars were grown in Ellinghausen-McCullough-Johnson-Harris (EMJH) media supplemented with *Leptospira* Enrichment (Difco™ Leptospira Medium Base EMJH, and Enrichment, Becton, Dickinson and Company, Sparks, MD, USA). A 4–7-day-old culture of each of the serovars adjusted to 0.5 McFarland standard was used for testing. In a 96-well plate, 50 µL of the *Leptospira* culture was mixed with 50 µL of each of the AGM serum samples diluted to 1:25 in PBS. After 1.5 to 2 h of incubation at 29 °C, the samples were screened for the presence of agglutination reaction to each of the serovars using a darkfield microscope. All the samples with 50% agglutination were recorded as positive for the presence of antibodies. Positive controls were set up using homologous antiserum (Royal Tropical Institute, KIT Biomedical Research, Amsterdam, The Netherlands) to each of the serovars included in the panel. Initially, all the serum samples at a final dilution of 1:50 were screened for the presence of agglutinating antibodies. Endpoint titers were determined for each of the positive samples against positive serovars.

DNA was extracted from the supernatant of homogenized AGM kidney samples using DNeasy Blood & Tissue Kits (QIAGEN Scientific Inc., Germantown, MD, USA), following the manufacturer's protocol. TaqMan probe-based Real-time PCR (RT-PCR) was performed on a Smart Cycler (Cepheid Inc., Sunnyvale, CA, USA) as described previously [26]. The *lipL32* gene present in pathogenic members of the genus *Leptospira* was targeted. A cycle threshold value above 40 cycles was considered negative. DNA extracted from a positive *Leptospira* culture served as positive control and sterile water served as the negative control.

Both kidneys, when available, were collected and fixed in 10% neutral buffered formalin, trimmed, processed, and stained with hematoxylin and eosin (H & E). Two board-certified veterinary pathologists evaluated renal histopathology. The association between MAT results and interstitial nephritis was compared using Fisher's exact test in GraphPadPrism (GraphPad Software 8.30, La Jolla, CA, USA). The statistical significance was set at $p = 0.05$.

5. Conclusions

Our study suggests that AGMs may serve as asymptomatic reservoirs, play a role in environmental contamination, and contribute to *Leptospira* transmission to humans and animals. The impact of *Leptospira* infection in renal pathology in AGMs warrants further investigation. AGMs residing in a natural setting in an insular, surface-limited, *Leptospira* endemic, geographic region may offer opportunities for comparative studies to advance the field of leptospirosis.

Author Contributions: Conceptualization, S.R.; formal analysis, S.R., K.S.; funding acquisition, S.R.; investigation, S.R., P.B., K.S., A.B., R.B.; methodology, S.R., P.B., K.S.; project administration, S.R.; resources, S.R., A.B., and R.P.; supervision, S.R.; writing—original draft, S.R., P.B.; writing—review and editing, S.R., P.B., K.S., A.B., R.P. All authors have read and agreed to the published version of the manuscript.

Funding: This research received no external funding.

Acknowledgments: We thank Shamara Welcome for technical assistance with MAT, and Katalina Cruz and Jennifer Ketzis for sharing kidney samples from their project. The project was funded by the One Health Center for Zoonosis and Tropical Medicine at the Ross University School of Veterinary Medicine.

Conflicts of Interest: The authors declare no conflict of interest.

References

1. Haake, D.A.; Levett, P.N. Leptospirosis in humans. *Curr. Top. Microbiol. Immunol.* **2015**, *387*, 65–97.
2. Ellis, W.A. Animal Leptospirosis. *Curr. Top. Microbiol. Immunol.* **2015**, *387*, 99–137. [PubMed]
3. Poingt, J.-P.; Perolat, P.; Jouaneau, C.; Gysin, J.; Vie, J.-C.; Baranton, G. Occurrence of Severe Leptospirosis in a Breeding Colony of Squirrel Monkeys. *Am. J. Trop. Med. Hyg.* **1992**, *46*, 538–545.
4. Szonyi, B.; Agudelo-Flórez, P.; Ramírez, M.; Moreno, N.; Ko, A.I. An outbreak of severe leptospirosis in capuchin (Cebus) monkeys. *Vet. J.* **2011**, *188*, 237–239. [CrossRef]
5. Gonzalez Astudillo, V.; Wehdeking Hernandez, D.; Pena Stadlin, J.; Arias Bernal, L.; Lombo Rodriguez, D.A.; Astudillo Hernandez, M. comparative seroprevalence of leptospira interrogans in colombian mammals along a climatic gradient. *J. Zoo Wildl. Med.* **2012**, *43*, 768–775. [CrossRef] [PubMed]
6. Asai, T.; Kinjo, T.; Minamoto, N.; Sugiyama, M.; Matsubayashi, N.; Narama, I. Prevalence Of Antibodies To Five Selected Zoonosis Agents In Monkeys. *J. Vet. Med. Sci.* **1991**, *53*, 553–559. [CrossRef] [PubMed]
7. Marshall, R.B.; BAskerville, A.; Hambleton, P.; Adams, G.D.J. Benign leptospirosis: The pathology of experimental infection of monkeys with Leptospira interrogans serovars balcanica and tarassovi. *Br. J. Exp. Pathol.* **1980**, *61*, 124–131.
8. Hambleton, P.; Baskerville, A.; Marshall, R.B.; Harris-Smith, P.W.; Adams, G.D. Metabolic sequelae of experimental leptospirosis in grivet monkeys. *Br. J. Exp. Pathol.* **1980**, *61*, 124–131.
9. Marchevsky, R.S.; Lenzi, H.L.; Pinto, M.A.; Machado, M.P.; Da Silva, M.F.; Da Silva, J.J.P.; Pereira, M.M. Experimental Leptospirosis In Marmoset Monkeys (*Callithrix jacchus*): A New Model For Studies Of Severe Pulmonary Leptospirosis. *Am. J. Trop. Med. Hyg.* **2005**, *72*, 13–20.
10. Rajeev, S.; Conan, A.; Pratt, N.; Beierschmitt, A.; Palmour, R. High Leptospira seroprevalence in captive and wild-caught vervet monkeys (*Chlorocebus sabeus*) on the Caribbean island of Saint Kitts. *J. Vet. Diagn. Investig.* **2017**, *29*, 930–934. [CrossRef]
11. Costa, F.; Hagan, J.E.; Calcagno, J.; Kane, M.; Torgerson, P.; Martinez-Silveira, M.S.; Stein, C.; Abela-Ridder, B.; Ko, A.I. Global Morbidity and Mortality of Leptospirosis: A Systematic Review. *PLoS Negl. Trop. Dis.* **2015**, *9*, 1. [CrossRef] [PubMed]
12. Shiokawa, K.; Llanes, A.; Hindoyan, A.; Cruz-Martinez, L.; Welcome, S.; Rajeev, S. Peridomestic small Indian mongoose: An invasive species posing as potential zoonotic risk for leptospirosis in the Caribbean. *Acta Trop.* **2019**, *190*, 166–170. [CrossRef]
13. Rajeev, S.; Shiokawa, K.; Llanes, A.; Rajeev, M.; Restrepo, C.M.; Chin, R.; Cedeño, E.; Ellis, E. Detection and Characterization of Leptospira Infection and Exposure in Rats on the Caribbean Island of Saint Kitts. *Animals* **2020**, *10*, 350. [CrossRef] [PubMed]
14. Shiokawa, K.; Welcome, S.; Kenig, M.; Lim, B.; Rajeev, S. Epidemiology of Leptospira infection in livestock species in Saint Kitts. *Trop. Anim. Health Prod.* **2019**, *51*, 1645–1650. [CrossRef] [PubMed]
15. Larson, C.R.; Dennis, M.; Nair, R.V.; Llanes, A.; Peda, A.; Welcome, S.; Rajeev, S. Isolation and characterization of Leptospira interrogans serovar Copenhageni from a dog from Saint Kitts. *JMM Case Rep.* **2017**, *4*, e005120. [CrossRef]
16. Pratt, N.; Conan, A.; Rajeev, S. Leptospira Seroprevalence in Domestic Dogs and Cats on the Caribbean Island of Saint Kitts. *Vet. Med. Int.* **2017**, *2017*, 1–6. [CrossRef]
17. McGuire, M.T. The St. Kitts vervet (*Cercopithecus aethiops*). *J. Med. Primatol.* **1974**, *3*, 287–295. [CrossRef]
18. LaCroix, K.; Callanan, J.J.; Cruz-Martinez, L.; Rajeev, S. Prevalence of Renal Lesions in the Small Indian Mongoose (*Herpestes auropunctatus*) Inhabiting the Caribbean Island of Saint Kitts. *J. Wildl. Dis.* **2018**, *54*, 881–884. [CrossRef]
19. Desai, S.; van Treeck, U.; Lierz, M.; Espelage, W.; Zota, L.; Sarbu, A.; Czerwinski, M.; Sadkowska-Todys, M.; Avdicová, M.; Reetz, J.; et al. Resurgence of Field Fever in a Temperate Country: An Epidemic of Leptospirosis among Seasonal Strawberry Harvesters in Germany in 2007. *Clin. Infect. Dis.* **2009**, *48*, 691–697. [CrossRef]
20. Katelaris, A.L.; Glasgow, K.; Lawrence, K.; Corben, P.; Zheng, A.; Sumithra, S.; Turahui, J.; Terry, J.; van den Berg, D.; Hennessy, D.; et al. Investigation and response to an outbreak of leptospirosis among raspberry workers in Australia, 2018. *Zoonoses Public Health* **2019**, *67*, 35–43. [CrossRef]
21. Levett, P.N. Usefulness of Serologic Analysis as a Predictor of the Infecting Serovar in Patients with Severe Leptospirosis. *Clin. Infect. Dis.* **2003**, *36*, 447–452. [CrossRef] [PubMed]

22. Yang, H.Y.; Hung, C.C.; Liu, S.H.; Guo, Y.G.; Chen, Y.C.; Ko, Y.C.; Huang, C.T.; Chou, L.F.; Tian, Y.C.; Chang, M.Y.; et al. Overlooked Risk for Chronic Kidney Disease after Leptospiral Infection: A Population-Based Survey and Epidemiological Cohort Evidence. *PLoS Negl. Trop. Dis.* **2015**, *9*, 1–15. [CrossRef]
23. Yang, C.W. Leptospirosis Renal Disease: Emerging Culprit of Chronic Kidney Disease Unknown Etiology. *Nephron* **2017**, *138*, 129–136. [CrossRef] [PubMed]
24. Jasinska, A.J.; Schmitt, C.A.; Service, S.K.; Cantor, R.M.; Dewar, K.; Jentsch, J.D.; Kaplan, J.R.; Turner, T.R.; Warren, W.C.; Weinstock, G.M.; et al. Systems Biology of the Vervet Monkey. *ILAR J.* **2013**, *54*, 122–143. [CrossRef] [PubMed]
25. Goris, M.G.A.; Hartskeerl, R.A. Leptospirosis serodiagnosis by the microscopic agglutination test. *Curr. Protoc. Microbiol.* **2014**, *32*, 1–18.
26. Stoddard, R.A. Detection of pathogenic leptospira spp. through real-time PCR (qPCR) targeting the lipL32 gene. *Methods Mol. Biol.* **2013**, *943*, 257–266.

© 2020 by the authors. Licensee MDPI, Basel, Switzerland. This article is an open access article distributed under the terms and conditions of the Creative Commons Attribution (CC BY) license (http://creativecommons.org/licenses/by/4.0/).

pathogens

MDPI

Article

Environmental Risk of Leptospirosis in Animals: The Case of the Republic of Sakha (Yakutia), Russian Federation

Olga I. Zakharova [1,*], Fedor I. Korennoy [1,2,*], Nadezhda N. Toropova [1], Olga A. Burova [1] and Andrey A. Blokhin [1]

1 Federal Research Center for Virology and Microbiology, Nizhny Novgorod Research Veterinary Institute-Branch of Federal Research Center for Virology and Microbiology, 603950 Nizhny Novgorod, Russia; 79875409639@yandex.ru (N.N.T.); burovaolga@list.ru (O.A.B.); and.bloxin2010@yandex.ru (A.A.B.)

2 Federal Center for Animal Health (FGBI ARRIAH), 600901 Vladimir, Russia

* Correspondence: olenka.zakharova.1976@list.ru (O.I.Z.); korennoy@arriah.ru (F.I.K.); Tel.: +7-950-617-16-01 (O.I.Z.)

Received: 27 April 2020; Accepted: 21 June 2020; Published: 23 June 2020

Abstract: Leptospirosis is a zoonotic natural focal disease caused by the pathogenic bacteria Leptospira. Its spread is related to certain ecological factors. The aim of the current research was to assess potential exposure to the infection as a function of environmental determinants in the Republic of Sakha (Yakutia), Russian Federation. We applied environmental niche modeling using leptospirosis cases in livestock and wild animals in 1995–2019 with regard to a set of landscape, climatic, and socioeconomic variables, both for the current climate and for the projected climate for 2041–2060. The MaxEnt model performed well (AUC = 0.930), with the mean temperature of the warmest quarter, mean diurnal range, land cover type, and altitude being the most contributing variables. Consequent zoning based on the proportion of high-risk cells within each administrative unit suggested that five out of the 36 districts of the Republic are at high risk in the current climate conditions, with three more districts expected to demonstrate a high risk by 2060. This study presents the first-ever attempt at leptospirosis ecological modeling in Russia. Its results correspond well to the findings of other authors and underline the importance of considering ecological factors when conducting a leptospirosis risk assessment.

Keywords: leptospirosis; MaxEnt; risk map; Republic of Sakha (Yakutia); climate change

1. Introduction

Leptospirosis is an infectious natural focal disease caused by the pathogenic bacteria *Leptospira* [1,2]. Currently, it is one of the most widely spread and neglected spirochetal zoonoses [3–11]. The pathogenic *Leptospira*, etiological agents of leptospirosis, cause at least 1,000,000 cases of the disease in humans and about 60,000 deaths annually [4]. Leptospirosis can be transmitted directly or indirectly from animals to humans, while human-to-human and human-to-animal transmission occur very rarely [5–16]. Many animal species, both domestic and wild, are affected. It has been proven that infected farm animals are responsible for the infection of humans, being a source of the agent of the disease. The pathogens of leptospirosis are microorganisms of the genus *Leptospira*, belonging to the independent family, *Leptospiraceae*, of the order *Spirochaetales*. The genus, *Leptospira*, combines 64 species and more than 300 serovars. Leptospirosis in animals is only caused by two subclades: the pathogens (which includes 13 species) and the intermediates (11 species) [6–8]. More recently, animal leptospirosis has attracted increased attention due to the large number of outbreaks in humans worldwide [9]. WHO has identified leptospirosis as a tropical disease of global importance requiring further research to understand its epidemiology, ecology, and pathology [4,10].

The animals infected with *Leptospira* contaminate the environment via the excretion of bacteria through their urine, which represents an indirect route of leptospirosis transmission. A pathogen transmission occurs either directly through a susceptible animal's contact with infected urine and other body fluids of an infected animal or via *Leptospira*-contaminated soil or water [11–13]. Historically, leptospirosis was known as an environment-borne infection, even before its etiological agent was identified [14–16] and soils in endemic areas were treated as its "environmental reservoirs" [17–20].

The incidence of leptospirosis varies in space and time and is closely related to climatic, ecological, and local socioeconomic factors [14,15,21]. A higher incidence is observed in tropical, humid, and temperate regions, especially during the rainy seasons, both in urban and rural areas [22,23], while research dedicated to leptospirosis patterns in northern regions are rare.

Based on the available data, it can be concluded that when assessing the transmission of leptospirosis to a susceptible livestock population, the geographical scale should be taken into account, in addition to global risk factors, in order to assess the local environmental and socioeconomic factors affecting animal infection [24]. Typically, massive outbreaks of leptospirosis in urban settings are associated with the abundance of rodents, being asymptomatic carriers of leprospires. In rural areas, outbreaks are linked to agricultural processes, such as animal breeding, while seasonal peaks exist in most affected areas suggesting that numerous environmental factors, such as torrential rains, floods, and cyclones may act as "triggers" of the disease outbreaks [25–35]. Environmental changes in human habitat conditions and urbanization combined with climate changes are the most important risk factors in the occurrence and spread of present and future leptospirosis outbreaks [18,36,37].

The complexity of pathogen transmission in leptospirosis is a serious problem for determining control strategies, especially in endemic areas. Epidemiological surveillance of the affected territories is the main and central step in the sequence of measures aimed at the rehabilitation of those territories after a leptospirosis outbreak [37–39]. Therefore, there is a need to select precise and cost-effective tools to improve existing surveillance procedures and strengthen control strategies. Geographic information systems (GIS) and geospatial statistics are now significantly improved and widely used in public health research. They can improve epidemiology and disease control measures. The goal of this study was the predictive modeling of the leptospirosis risk zones in the Republic of Sakha (Yakutia), Russian Federation (hereinafter RSY), in connection with a set of environmental and socioeconomic factors using the method of ecological niche modeling, and forecasting changes in risk levels due to the expected climate changes for the period up to 2060. To achieve this goal, the following objectives were formulated:

1. To summarize the epidemiological data on the occurrence of animal leptospirosis in the RSY for 1995–2019;
2. To identify the main environmental and socioeconomic factors contributing to the occurrence of animal leptospirosis outbreaks in the RSY;
3. To develop an environmental suitability based leptospirosis risk map using the MaxEnt ecological niche model, subsequently averaging the risks within municipal areas to provide zoning in accordance with the currently applied veterinary practice.

2. Results

2.1. Epidemiological Analysis

From 1995 to 2019, 434 outbreaks of leptospirosis in animals were recorded in the RSY (Figure 1). In total, for this period, 2728 animals were infected, of which 31.4% were cattle (859 animals), and 57.2% were horses (1561 animals). The proportion of infected pigs and sheep was 7.6% and 1.2%, respectively. The rest of the infected animals were domestic (cats, dogs), and wild animals (reindeers, rabbits, etc.), which together sum up to 2.6% of all infected animals. The yearly incidence of leptospirosis in livestock species (cattle, horses, pigs, and sheep) in the RSY for the period from 1995 to 2019 is presented in Figure 2. Based on the data of recorded cases for this period, most outbreaks were noted in 2008. In total, there were three periods during which an increase in the incidence could be observed: (1) from 1995 to 2008; (2) from 2014 to 2016; and (3) from 2017 to 2018. The seasonality diagram suggests that a peak of the disease incidence was observed in the warm season (Figure 3). The incidence of leptospirosis reached its highest value in March and June, amounting to 15.9% and 15.4% of all outbreaks, respectively. In cattle, the maximum increase in incidence occurs in the spring–summer period: starting in March (12.2%) and April (17.8%). The peak incidence in cattle was in June (27.3%). In horses, the peak incidence was observed also in the spring–summer period, but the maximum incidence was found in both March (17.37%) and June (14.55%). In pigs, the seasonality pattern was different from that in cattle and horses. The incidence peaks in pigs were observed in January, February, April, June, and September. This is presumably related to the return of the Leptospira hosts: mice, voles, and rats in pig farms. The incidence of cattle and horses may be determined by the timing of the grazing period, which in the RSY normally lasts from June to September for cattle and almost year-round for horses.

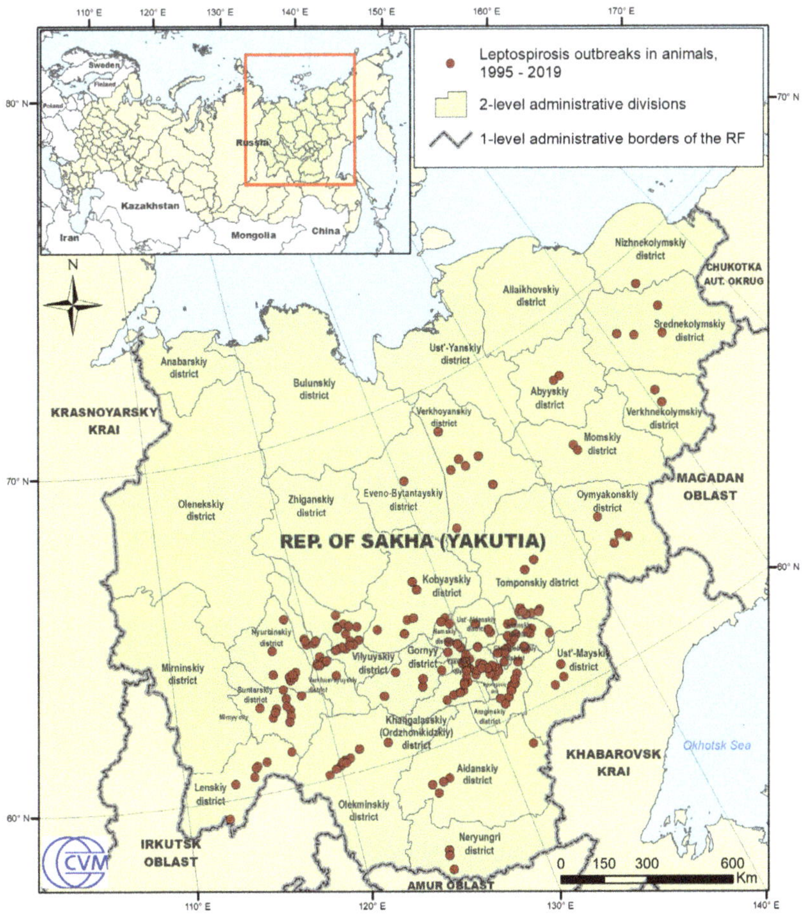

Figure 1. The distribution of recorded leptospirosis outbreaks in animals in the Republic of Sakha (Yacutia), 1995–2019.

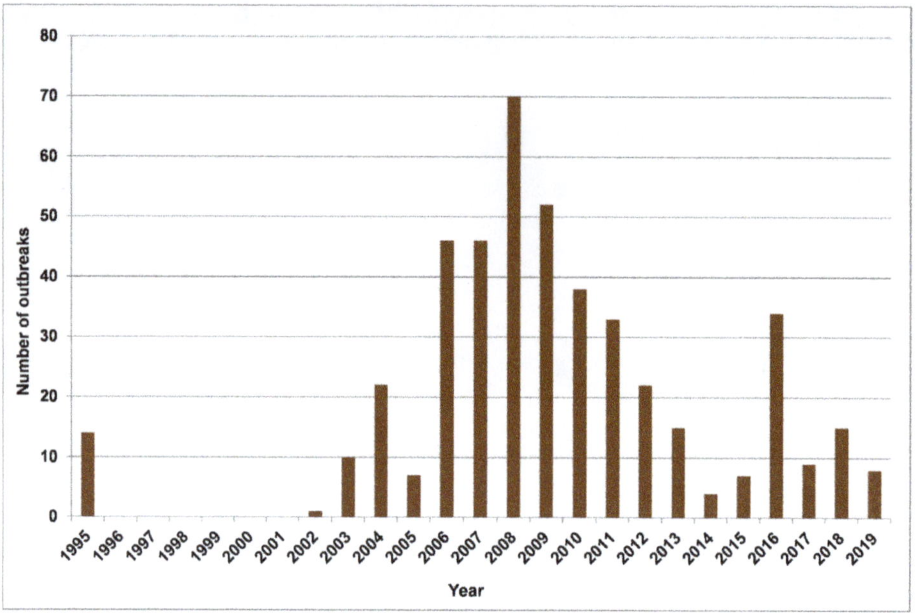

Figure 2. Yearly incidence of leptospirosis in livestock and wild animals in the Republic of Sakha (Yacutia), 1995–2019.

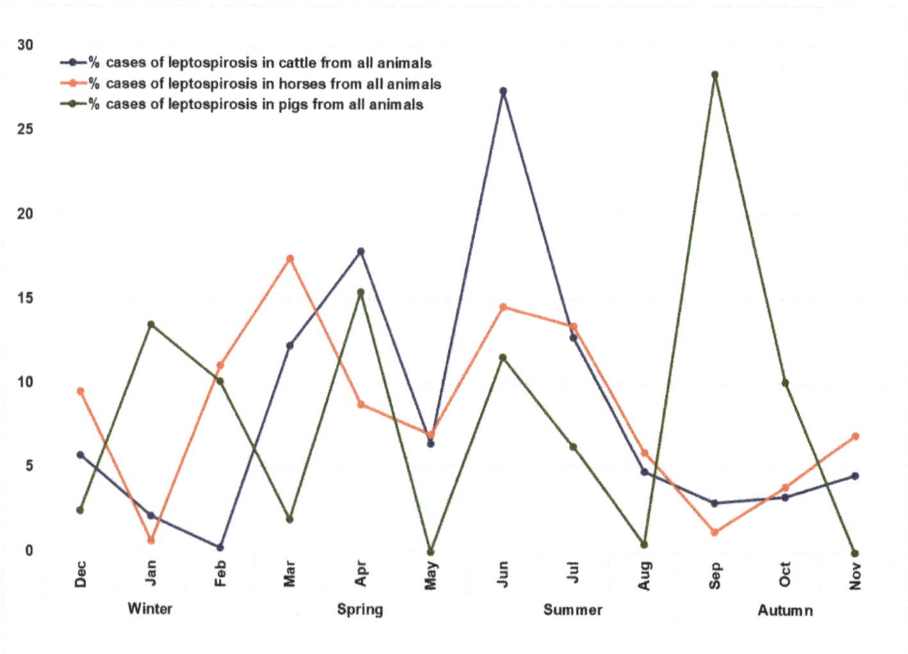

Figure 3. Seasonality of leptospirosis outbreaks in cattle, horses and pigs in the Republic of Sakha (Yacutia), 1995–2019.

2.2. Environmental Niche Modeling

An environmental niche model, implemented with MaxEnt, demonstrated a good ability to distinguish between true presence and pseudoabsence data (AUC = 0.930 ± 0.026). The most important environmental risk factors related to the leptospirosis presence points in the RSY were: mean temperature of the wettest quarter (Bio_8), altitude (alt), land cover, mean diurnal range (bio_2), temperature seasonality (bio_4), precipitation of the wettest month (bio_13), cattle density, and soil pH, and the importance of the permutations was 10.3%, 36.3%, 10.8%, 13.2%, 5.2%, 6.3% 5.0%, and 4.1%, respectively. The importance of these predictors was also confirmed by Jackknife analysis. The response curves, indicating the model gain using only a particular variable, are presented in Figure 4 for the eight most important variables (listed above). The curves suggest a maximum suitability to leptospirosis transmission in areas with the mean temperature of the wettest quarter being approximately 14 °C and above, low altitudes (less than 500 m above sea level), considerable variation in the monthly temperatures within a year (10 °C and above), pronounced seasonal temperature deviation of about ±20 °C, relatively low maximum precipitation of ~20–60 mm/month, relatively high cattle population density (more than 0.5 heads/km^2), and soil pH of 5.5 and above. The land cover categories most closely associated with the risk of leptospirosis were: Urban and Built-Up areas, Open Deciduous Needleleaf Forest, Open ground and rock outcrops, Coastal vegetation, Permanent Wetlands, and Grasslands. The soil units corresponding to the most suitable areas were Histic Gleysols Dystric, Gleyic Albeluvisols Abruptic, Haplic Podzols, and Haplic Cambisols Eutric. The rest of the variables used for modeling did not demonstrate a significant contribution to the model gain, while still providing a reasonable influence on the probability of having a leptospirosis outbreak. Lower maximum green vegetation fraction values were more suitable, showing a monotonic decrease over the study range. The distance to the nearest water body appears to have a negative relation with leptospirosis locations, suggesting that most of the recorded cases occurred not farther than 2500 m from water.

Table 1. Environmental and socioeconomic factors used as explanatory variables in an ecological niche model after removing correlated variables and variables with a high VIF.

Variable Name	Variable Description	Units	Data Type
Alt	Altitude	Meters	
Bio_2	Mean Diurnal Range (Mean of monthly (max temp–min temp))	°C × 10	Continuous
Bio_4	Temperature Seasonality (standard deviation × 100)	°C × 1000	Continuous
Bio_8	Mean Temperature of Wettest Quarter	°C × 10	Continuous
Bio_9	Mean Temperature of Driest Quarter	°C × 10	Continuous
Bio_13	Precipitation of Wettest Month	Millimeters	Continuous
Bio_14	Precipitation of Driest Month	Millimeters	Continuous
Bio_15	Precipitation Seasonality (Coefficient of Variation)	Proportion	Continuous
MGVF	Maximum Green Vegetation Fraction	Proportion	Continuous
Land cover	Land cover type	Land cover categories (see Table A1)	Categorical
Soils	Soil type	Soil categories (see Table A2)	Categorical
Soil pH	Soil pH at zero depth	pH × 10	Continuous
Water distance	Euclidean distance to the nearest freshwater body	Meters	Continuous
Cattle density	Density of cattle	Heads/km^2	Continuous

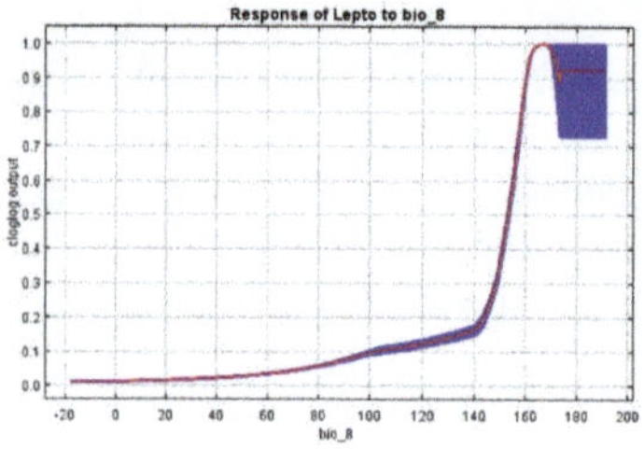

(**a**) X-axis: °C × 10

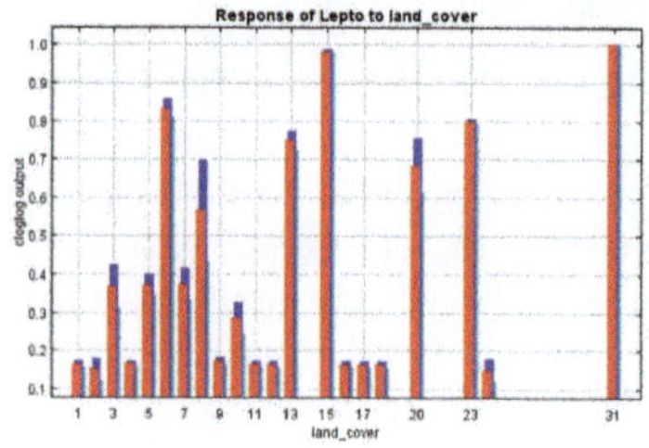

(**b**) X-axis: land cover categories (see Table A2)

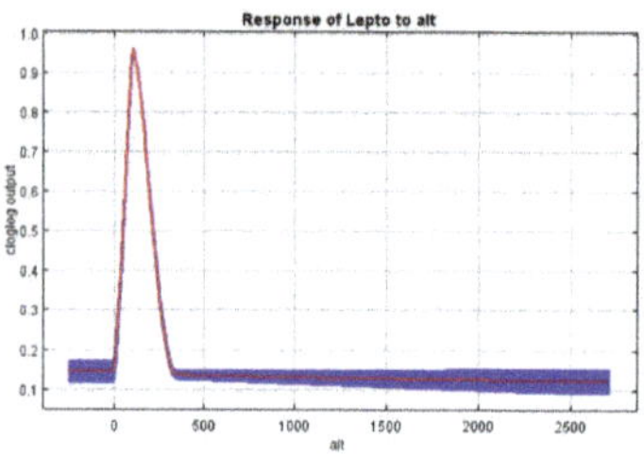

(**c**) X-axis: meters

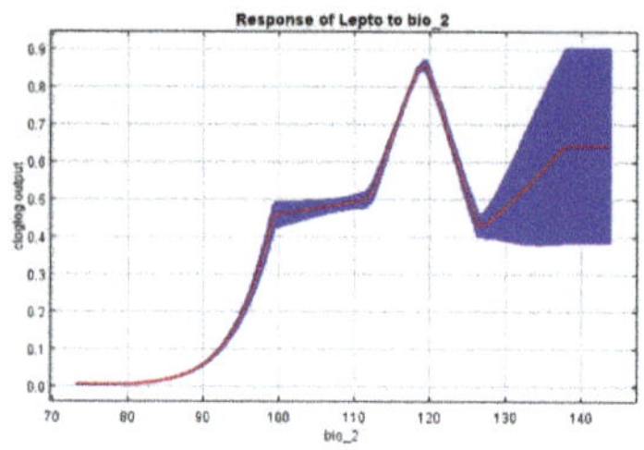

(**d**) X-axis: °C × 10

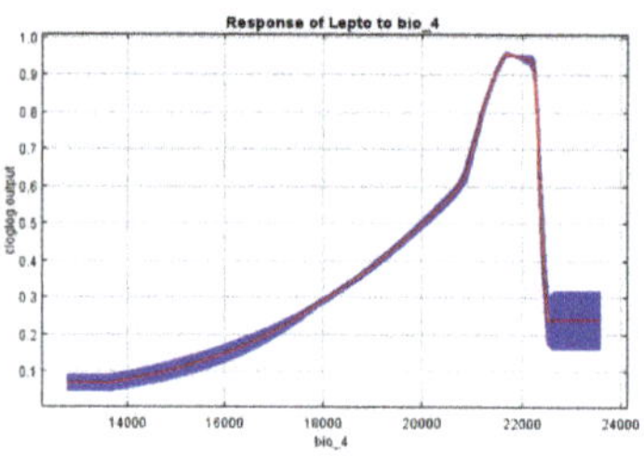

(**e**) X-axis: °C × 1000

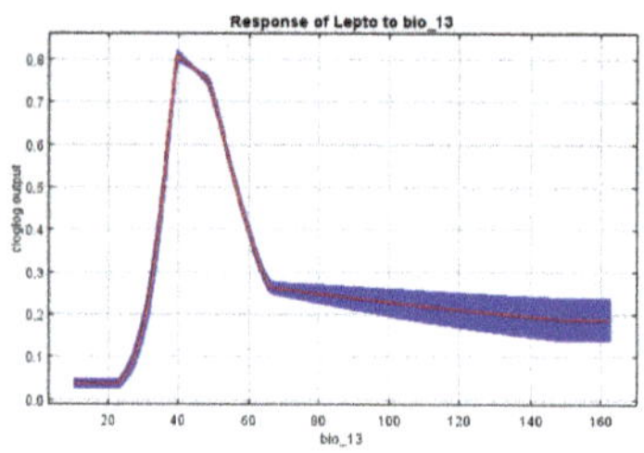

(**f**) X-axis: millimeters

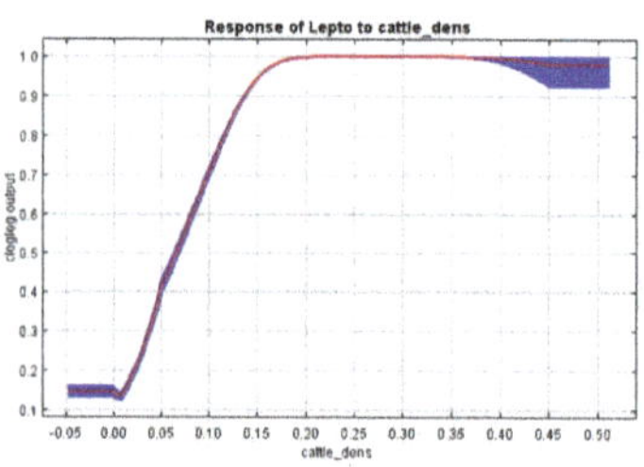

(**g**) X-axis: heads/km^2

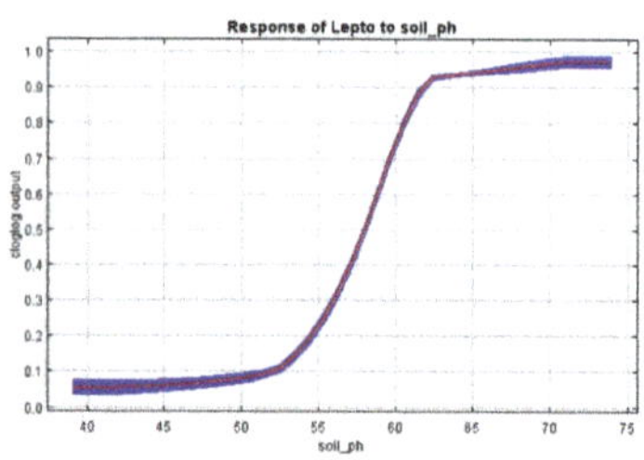

(**h**) X-axis: pH × 10

Figure 4. Response curves for 8 variables that contribute the most: (**a**) Bio_8—Mean Temperature of Wettest Quarter; (**b**) land cover type; (**c**) altitude; (**d**) bio_2—Mean Diurnal Range; (**e**) bio_4—Temperature Seasonality; (**f**) bio_13—Precipitation of Wettest Month; (**g**) cattle density; (**h**) soil pH. Red lines indicate average trends over 10 model replications while blue areas denote the standard deviation limits. The *Y*-axis is relative suitability calculated only using the particular variable. The *X*-axis variables' units are presented at a footnote of each diagram as well as listed in Table 1.

Suitability maps for the current and future climate are shown in Figure 5a,b, respectively, while Figure 5c shows a map of statistically significant suitability changes in the future, as compared to the current conditions. The maps clearly demonstrate that the most suitable areas are present in the south-eastern part of the RSY, which is also the most inhabited territory of the Republic, and extend in the direction of major rivers.

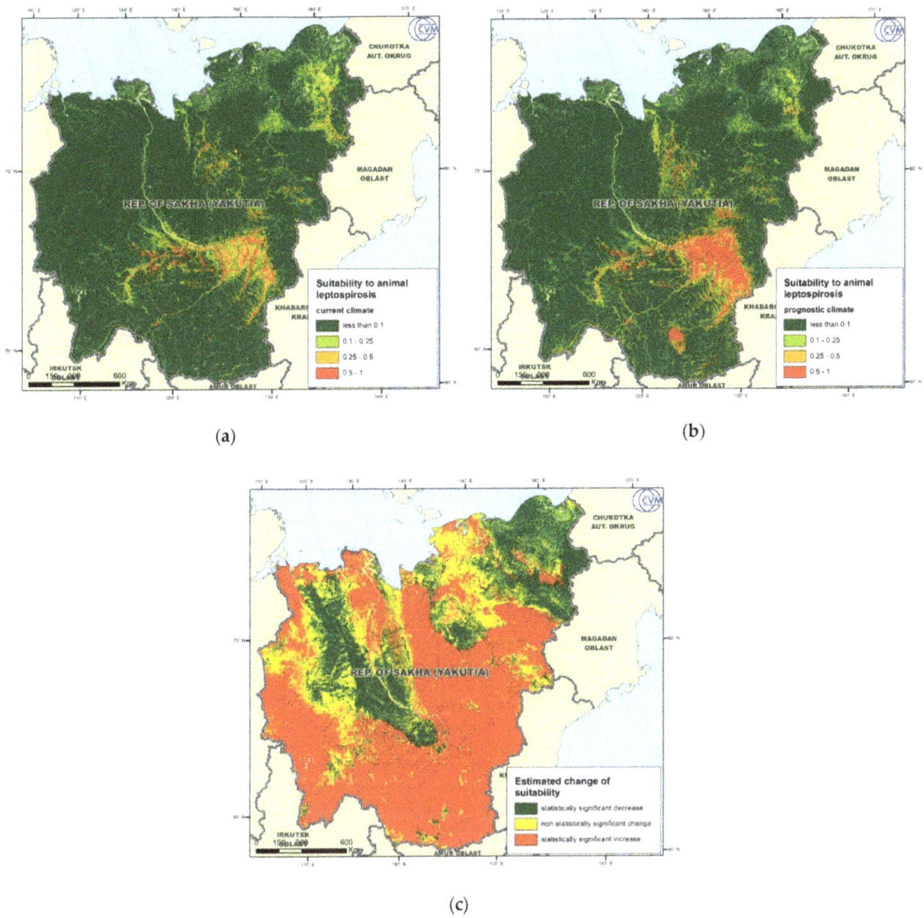

Figure 5. MaxEnt-derived gridded suitability maps for the current climate (**a**), projected climate (**b**), and statistically significant changes (**c**).

2.3. *Leptospirosis Risk Mapping*

Gridded suitability maps were converted into categorical zoning maps based on the proportion of high-risk cells within each administrative district. The risk maps for the current climatic conditions (Figure 6a) and for the projected climate (Figure 6b) suggest the existence of high- and medium-risk zones in the south-eastern part of the RSY. In the current climate, the high-risk zone includes: Namsky, Ust-Aldansky, Tattinsky, Alekseevsky, Churapchinsky districts, and the city of Yakutsk. In addition, 14 districts can be rated as medium-risk areas. Under the projected climate, the Medino-Kangalassky, Amginsky, and Ust-May districts may fall into the high-risk zone, and the medium-risk zone will increase by three districts, providing an expansion of leptospirosis into nearly half of the RSY territory.

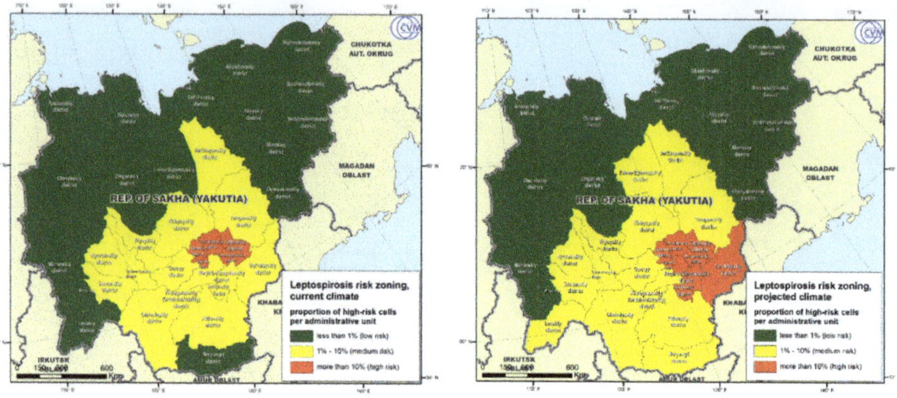

(a) (b)

Figure 6. Republic of Sakha (Yakutia) Leptospirosis risk zoning for the current climate (**a**) and for the projected climate (2041–2060) (**b**).

3. Discussion

In this study, for the first time, we collected data on leptospirosis outbreaks in the subarctic region of Russia from 1995 to 2019. We identified a number of environmental and socioeconomic factors that are associated with the disease's geographical spread. We used the ecological niche modeling (ENM) approach to better understand the relationship between the distribution of leptospirosis outbreaks and environmental variables and to construct predictive maps that illustrate the RSY territory suitability to the leptospirosis emergence.

The results of many studies have shown the importance of accounting for environmental factors when studying the epidemiology of leptospirosis [21,40,41].

Environmental factors play a significant role in the emergence of natural focal diseases, directly affecting pathogenic microorganisms or affecting the distribution and number of hosts and carriers of diseases and forming favorable conditions for the persistence and transmission of the disease [21]. Since the ENM tools were first used to predict outbreaks of leptospirosis in the subarctic region of Russia, we tested all commonly used environmental variables as potential predictors. While not directly comparable with the results of other authors due to the use of different variables, our findings reveal very similar distribution patterns of recorded leptospirosis outbreaks. Our choice of final variables was ultimately determined by the procedure of reducing multicollinearity, potentially leaving overboard some more general parameters, e.g., the yearly mean temperature and precipitation. Our results rely on more specific parameters, the interpretation of which may be difficult to compare with the results of other studies. Nevertheless, the revealed patterns may be treated as very similar to the ones reported in [21,40,41]. The mean temperature of the wettest quarter, temperature seasonality, temperature variation, and soil pH were the most important environmental factors determining the geographical spread of leptospirosis in the Republic of Sakha (Yakutia). In general, the combination of these factors suggests that the persistence and transmission of *Leptospira* favor humid areas with a considerable temperature variation throughout the year. Since leptospirosis is an infectious disease that can be transmitted through water, we used the Euclidean distance to nearest water sources for the transmission of leptospirosis indirectly. Unfortunately, the corresponding variable does not significantly affect the geographical spread of leptospirosis, while demonstrating an expected pattern of the suitability decrease as the distance to water increases. From a socioeconomical perspective, only the cattle population density was used, which was found to be statistically significantly associated with the recorded outbreaks. This factor may be treated as a proxy of human presence and economic activity in the sparsely populated area of the RSY, as well as an indicator of the cattle population's exposure to leptospires.

There is a lot of evidence that virulent *Leptospira* can survive and remain infectious in an environment for several months, particularly in soils [42,43]. According to studies of other authors, leptospires are more often isolated from soils, rather than from freshwater samples [44–46]. These data confirm the hypothesis that soils can be a reservoir of *Leptospira* spp. and specifically of its pathogenic strains [47,48]. Due to the climatic conditions, the soils in the RSY are in the frozen state for most of the year. This may explain the fact that most of the RSY has a low leptospirosis suitability, while an elevated suitability was found in proximity to rivers and populated places that occupy the most naturally favorable areas of the Republic. The soil units that demonstrated most significant associations with leptospirosis outbreaks can be described as tundra and taiga-specific soils, with a relatively high organic content. Soil pH was found to more significantly influence suitability to leptospirosis, with higher values being more preferable, which correspond well to the findings in [40,41].

The combination of low altitudes and specific land cover categories associated with the known locations of leptospirosis outbreaks may either suggest a data bias, related to the georeferencing of outbreaks to the nearest settlement, or point to the disease emergence in anthropogenic landscapes, where the transmission of the pathogen to farm animals and humans is possible. To reduce a potential bias of data towards the populated areas, we introduced a "data bias correction factor" to the MaxEnt model, which was the density of settlements. Such a factor assumes that the input presence locations might be primarily recorded in close proximity to populated places and corrects the resulted suitability distribution accordingly. The observed pattern of the suitability response to the maximum vegetation fraction with lower maximum green vegetation fraction (MGVF) values being more suitable may also suggest a primary disease transmission within inhabited areas, which are less covered by vegetation.

The predicted distribution of the suitability change in a future climate suggests an increase of leptospirosis risk over the large area of the RSY. Some areas in the north and northeast of the Republic, however, demonstrate a statistically significant reduction of suitability. The observed pattern looks quite similar to the estimated change of the Bio_8 variable that stays for the Mean Temperature of the Wettest Quarter. Considering an expected temperature rise over the entire territory of the RSY, one can assume a wettest season's shift towards colder months that will lead to the decrease of the mean temperature. Those areas of expected suitability reduction mainly correspond to lowlands surrounded by mountainous areas, such as Verkhoyansk and Chersky Ranges. Such territories are characterized by the stagnation of cold air masses flown down from mountains and thus normally demonstrate relatively lower air temperatures than flat or highland areas.

The results obtained allow optimizing the long-term control and monitoring program aimed at leptospirosis prevention in high-risk areas that, in particular, includes compulsory vaccination of domestic animals.

4. Materials and Methods

4.1. Study Area

The Republic of Sakha (Yakutia) (RSY) is one of the 85 federal subjects of the Russian Federation, located in its north-eastern part. Its mainland extends from 55° to 74° North and from 105° to 163° East (Figure 1). With an area of more than 3,000,000 km^2, the Republic is the largest federal subject of the Russian Federation, as well as the largest administrative unit in the world. In addition, the population density is only about 0.3 persons/km^2, which is one of the lowest rates in the Russian Federation and in the world. More than one-third of the entire territory of the Republic is located beyond the Arctic Circle, more than 80% of the area is covered by forests, and more than two-thirds are mountains and plateaus. The Republic has a large number of natural water bodies (rivers and lakes). The climate ranges from continental in southern and central parts to subarctic and arctic in the north. The Republic is rich in minerals (coal, natural gas, diamonds, gold, etc.). The main inhabited zones are located along the major rivers, as well as around large subsoil deposit locations. The territory of the Republic is divided into 36 districts ("uluses"), which are second-level administrative units. The agricultural activity is mainly cattle farming, horse breeding, and reindeer husbandry.

4.2. *Leptospirosis Data*

The leptospirosis data were obtained from two official sources. The regional veterinary service reports provided information on the leptospirosis cases in farm animals. Such data are based on farm owners' claims followed by laboratory confirmation. Additionally, statistical veterinary reports of the Federal Service for Veterinary and Phytosanitary Surveillance (Rosselkhoznadzor) were used that contain the results on the positive testing of both farm and wild animals during the passive surveillance program being implemented in the RSY. The surveillance program implies testing of 25% of the susceptible farm animals with occasional testing of wild animals. According to the currently effective National Standard "GOST 25386-91", the laboratory confirmation was carried out by the Yakut Republican Veterinary Testing Laboratory via a microagglutination reaction test (MAT) using a set of 7 reference cultures: Pomona, Tarassovi, Canicola, Hebdomadis, Sejroe, Grippotyphosa, and Icterohaemorrhagiae. The reaction was evaluated in a positive cutoff, with a serum dilution of 1:50 for unvaccinated, and 1:100 for vaccinated animals with dark-field microscopy. For previously vaccinated animals, testing was conducted at least 3 months after the vaccination. In case of a positive MAT result and no clinical signs, confirmatory testing by real-time PCR was used.

In total, 454 outbreaks of leptospirosis were recorded in the RSY during the study period of 1995–2019 (Figure 1). In this study, we defined an outbreak as the reported occurrence of leptospirosis in a geographically localized animal population (herd on a certain pasture, farm, village, etc.) with at least one infected or seropositive animal, confirmed by laboratory research methods. The database contains outbreaks of leptospirosis detected in cattle, small ruminants, pigs, horses, dogs and cats, and wild animals, including rabbits, rats, and mice.

The following indicators, important for further modeling, were recorded for each outbreak: Geographical coordinates, exposed animal species, and their number, as well as the date the disease, were diagnosed. After excluding inaccurate records (in particular, the cases without geographical coordinates or with incorrect ones), the database contained 434 outbreaks of leptospirosis, in which 2728 out of the 46,569 susceptible animals were exposed. Taking into account the number of repeated outbreaks recorded at different time periods in one location, the total number of unique locations was 207. The map of outbreaks is shown in Figure 1. The geographic data are converted into a shapefile format for visualization and further modeling.

4.3. *Modeling Approach*

We used the method of ecological niche modeling with the principle of maximum entropy (MaxEnt) to identify the relationship of recorded outbreaks of leptospirosis with a complex of abiotic and biotic variables. This approach, described by Phillips [49], is currently one of the most popular methods for modeling the distribution of a studied phenomenon in relation to environmental factors based on presence-only data [50,51].

The principle underlying the maximum entropy method is to search for the probability distribution of each factor that would most uniformly describe its change within the range specified by the presence data (i.e., would have a maximum informational entropy) [50]. The resulting map shows the probability that a combination of environmental factors at each cell of the study area is suitable for the studied phenomenon. Predictive power is estimated by the ability of the model to distinguish between the presence and pseudoabsence data and is expressed by the area under the ROC curve (AUC). This indicator shows the probability that a randomly selected presence point will have a higher prognostic value than a randomly selected pseudoabsence point [52]. In addition, this method makes it possible to evaluate the contribution of each variable to the obtained model and to draw a conclusion about the most significant biological factors by means of the Jackknife technique, which is based on the sequential exclusion of each variable from the model and a comparison of the model gain with and without the variable, as well as with the variable only.

MaxEnt modeling was performed using cross-validation, with the presence data divided into 10 folds and each fold in turn used for testing. Summary results, thus, include average values over 10 replications as well as standard deviation limits. Within each replication, 5000 iterations were used to reach the highest gain with 0.00001 convergence threshold.

Using two sets of BIO variables (see Section 4.4.1 below), modeling was performed both for the current climatic conditions and for a projected climate for the period 2041–2060.

To account for a potential data bias related to a sampling distribution shifted towards populated places, we included a "data bias" parameter expressed as the settlement density. It introduces an assumption that incidence records are more likely to be made in or in close proximity to towns, settlements, and villages.

A comparison of the suitability for the current and projected climates was conducted by mapping statistically significant estimated changes for each cell of the study area. The map is developed by subtracting the current suitability value from the predicted suitability and then normalizing the difference with the standard deviation calculated for the current suitability. Three classes of expected changes were distinguished:

1. Difference within ±1 of the standard deviation was recognized as statistically insignificant;
2. Difference below −1 of the standard deviation indicates the places in which the suitability reduction was predicted;
3. Difference above +1 of the standard deviation indicates the places in which an increase in suitability was predicted.

To compare risks at the level of administrative divisions, zoning was performed by calculating the proportion of spatial cells with a suitability above 0.5 ("high-risk cells") for each municipal district, both for the current and projected climates, and, subsequently, categorizing the risk levels in three classes based on the distribution of the obtained proportions [53,54].

4.4. Explanatory Variables

4.4.1. Climatic Variables

We used a set of bioclimatic variables BIO as climatic factors for the current and projected climates (https://worldclim.org/data/index.html). This set consists of 19 variables (BIO_1 ... BIO_19), describing the peak and trend values based on the average monthly air temperatures and precipitation: (I) for 1970–2000, according to weather stations data; and (II) projected for the period 2041–2060, calculated on the basis of the downscaled INMCM4 climate model. This is the atmosphere-ocean global climate model developed by the Institute of Numerical Mathematics, Russian Academy of Science. The model has participated in the International Coupled Model Intercomparison Project, phase 5 (CMIP5) [55]. A modeling scenario was used that corresponds to a representative concentration pathway of RCP8.5. This scenario determines that the concentration of greenhouse gas and anthropogenic effects in the atmosphere raises in the toughest trajectory and presents the most stringent option for climate change. The initial spatial resolution of the data is 30 arc seconds.

4.4.2. Animal Host Density

Leptospirosis is detected mainly in those areas where people interact with animals, with excretions of infected animals or with contaminated environmental objects [11]. Rodents and some domestic animals, such as pigs and dogs, are considered the riskiest for the transmission of leptospires to humans [31]. Leptospirosis is considered an epidemic infection that can be transmitted directly to humans from water, soil, and urine-contaminated food from infected host animals [12]. We used the density of the cattle population as the main species of farm livestock in the study region. This indicator was used as a proxy for the intensity of animal husbandry and human agricultural activity, increasing the likelihood of contact between animals and wild species. Animal density data were obtained from the FAO geoportal (http://www.fao.org/geonetwork/srv/en/main.home) in the raster format.

4.4.3. Proximity to Water Bodies

The literature sources confirm that the pathogenic Leptospira is associated with the presence of water bodies, including rivers, streams, lakes, and springs. Therefore, the distance to natural and artificial water sources is a potential risk factor for leptospirosis [56].

The database of freshwater reservoirs (rivers, lakes) in the RSY was recovered from the vector dataset, "Digital model of Russia on a scale of 1:500,000" (https://www.esri-cis.ru/products/). The distance to the nearest water body was calculated using the Euclidean distance tool (Spatial Analyst, ArcMap).

4.4.4. Land Cover Type

As a source of data on the types of land cover in the Russian Federation, we used a digital map, created according to the Proba-V satellite system for the years 2000–2018, with an original spatial resolution of 100 × 100 m [57]. The categories of land cover presented within the entire database are listed in Table A1.

4.4.5. Vegetation Index

An average maximum green vegetation fraction (MGVF) for 2001–2012 was used as an indicator of the intensity of vegetation within the model area [58].

4.4.6. Soil Type

Data on the soil type in the model territory were obtained from the Unified State Register of Soil Resources of Russia (http://egrpr.esoil.ru/content/1DB.html), which has 255 soil units, in accordance with the world harmonized database of soil data (http://www.fao.org/soils-portal/soil-survey/soil-maps-and-databases/harmonized-world-soil-database-v12/en/). The soil units presented in the RSY are listed in Table A2. The original vector data were converted into a raster format.

4.4.7. Soil pH

To study the potential relationship of leptospirosis outbreaks with soil acidity, a variable was included in the model that displays the soil pH at zero depth, according to the ISRIC-World Soil Information. The data are presented in raster format, with an initial resolution of 250 × 250 m^2 [59].

4.5. Data Processing and Software

The Asia Lambert Conformal Conic map projection was used for mapping and geospatial analysis. All geospatial variables presented in raster format were reduced to a single spatial resolution of 1 × 1 km and converted to the ASCII format. A preliminary correlation analysis of all variables was performed to reduce the multicollinearity of the MaxEnt model using the "usdm" package in the R 3.6.2 software environment [60]. At the first step, of each pair of highly correlated variables (Pearson correlation coefficient above 0.85) the one was excluded with a larger VIF (Variance Inflation Factor). For the remaining variables, VIF was analyzed, and indicators with VIF > 10 were additionally excluded. The final set of predictors is presented in Table 1. Statistical data processing was carried out using the MS Office Excel package (Microsoft, Redmond, WA, USA). To process the geodata and create maps, the ArcMap 10.7.1 geographic information system (ESRI, Redlands, CA, USA) was used. Ecological modeling with the maximum entropy method was carried out using the MaxEnt software [61].

5. Conclusions

This article discusses, for the first time, the problem of the spread of leptospirosis in the Republic of Sakha (Yakutia), a subarctic region with a continental to arctic type of climate.

The results of this study demonstrate that the distribution of leptospirosis cases in the Republic is significantly influenced by complex landscape and climatic factors, which corresponds well to the conclusions made by other authors. The findings suggest an expansion of the leptospirosis risk zone in the Republic under the projected climate change in 2041–2060. The method of risk zoning may be recommended to the national veterinary service as a basis for conducting targeted surveillance of leptospirosis in the environment and in livestock populations.

Author Contributions: Conceptualization, O.I.Z., A.A.B., and F.I.K.; methodology, F.I.K. and O.I.Z.; software, F.I.K. and O.I.Z.; validation, A.A.B. and F.I.K.; formal analysis, F.I.K. and O.I.Z.; investigation, O.I.Z. and N.N.T.; resources, N.N.T. and O.A.B.; data curation, F.I.K.; writing—original draft preparation, O.I.Z. and F.I.K.; writing—review and editing, F.I.K., A.A.B. and O.I.Z.; visualization, O.I.Z. and F.I.K.; supervision, A.A.B.; project administration, A.A.B., F.I.K., and O.I.Z.; funding acquisition, A.I.B. All authors have read and agree to the published version of the manuscript.

Funding: The work was supported by the Federal Research Center for Virology and Microbiology for government assignment. The data were partly obtained by collaboration under a scientific project No. 18-05-60037 by Russian Foundation for Basic Research (RFBR).

Acknowledgments: We thank the coordinator of the epizootology and animal risk assessment group Andrei E. Gogin (Federal Research Center for Virology and Microbiology) for their valuable contribution to the discussion of our work.

Conflicts of Interest: The authors declare no conflicts of interest.

Appendix A

Table A1. Land cover types (adopted from [57]).

Category Number	Land Cover Type
0, 30	No data
1	Dark Evergreen Needleleaf Forest
2	Light Evergreen Needleleaf Forest
3	Broadleaf forest
4	Deciduous Needleleaf Forest
5	Evergreen Needleleaf shrubs
6, 7, 21, 22	Permanent Wetlands
8	Grasslands
9	Broadleaf shrubs
10	Coniferous mixed forests
11	Mixed forests
12	Deciduous mixed forests
13, 19	Open ground and rock outcrops
14	Steppe
15	Coastal vegetation
16	Shrubby tundra
17	Grassy tundra
18	Shrub tundra
20	Water
23	Open Deciduous Needleleaf Forest
24	Fresh Burns
31	Urban and Built-Up areas
32	Snow and Ice
33	Arable land

Table A2. Soil units presented in the Republic of Sakha (Yakutia) (adopted from [59]).

Category Number	Soil Unit According to WRB06 [1]	Soil Unit According to FAO88 [2]
3	Haplic Cryosols Eutric	Gelic Regosols
6	Haplic Cryosols Reductaquic	Gelic Gleysols
7	Haplic Cryosols Oxyaquic	Gelic Gleysols
12	Spodic Cryosols Dystric	Gelic Podzols
22	Histic Gleysols Dystric	Dystric Gleysols
24, 35	Gleyic Albeluvisols Abruptic	Gleyic Podzoluvisols
41–46	Umbric Albeluvisols Abruptic	Eutric Podzoluvisols
48	Gleyic Albeluvisols Abruptic	Gleyic Podzoluvisols
49, 51	Umbric Albeluvisols Abruptic	Eutric Podzoluvisols
54, 55	Gleyic Albeluvisols Abruptic	Gleyic Podzoluvisols
57	Carbic Podzols	Haplic Podzols
60	Histic Podzols	Haplic Podzols
61	Haplic Podzols	Haplic Podzols
66	Entic Podzols	Cambic Podzols
69, 70	Haplic Cambisols Dystric	Dystric Cambisols
71	Haplic Cambisols Humic	Humic Cambisols
81	Haplic Cambisols Dystric	Gelic Cambisols
83	Rendzic Leptosols Eutric	Rendzic Leptosols
97	Haplic Cambisols Eutric	Dystric Cambisols
99	Haplic Cambisols Eutric	Eutric Cambisols
106, 108, 109, 112	Greyic Phaeozems Albic	Haplic Greyzems

[1.] World reference base for soil resources 2006, FAO; [2.] FAO/UNESCO Soil Map of the World, 1988.

References

1. Breneva, N.V.; Balakhonov, S.V. Endemicity and enzooticity aspects of leptospirosis. *J. Microbiol. Epidemiol. Immunobiol.* **2019**. [CrossRef]
2. Ukhovskyi, V.V.; Vydayko, N.B.; Aliekseieva, G.B.; Bezymennyi, M.V.; Polupan, I.M.; Kolesnikova, I.P. Comparative analysis of incidence of leptospirosis among farm animals and humans in Ukraine. *Regul. Mech. Biosyst.* **2018**. [CrossRef]
3. Bolin, C. Leptospirosis. In *Emerging Diseases of Animals*; Brown, C., Bolin, C., Eds.; ASM Press: Washington, DC, USA, 2000; pp. 185–200. [CrossRef]
4. Costa, F.; Hagan, J.E.; Calcagno, J.; Kane, M.; Torgerson, P.; Martinez-Silveira, M.S.; Stein, C.; Abela-Ridder, B.; Ko, A.I. Global Morbidity and Mortality of Leptospirosis: A Systematic Review. *PLoS Negl. Trop. Dis.* **2015**, *9*. [CrossRef] [PubMed]
5. Allan, K.J.; Halliday, J.E.B.; Moseley, M.; Carter, R.W.; Ahmed, A.; Goris, M.G.A.; Hartskeerl, R.A.; Keyyu, J.; Kibona, T.; Maro, V.P.; et al. Assessment of animal hosts of pathogenic Leptospira in northern Tanzania. *PLoS Negl. Trop. Dis.* **2018**, *12*. [CrossRef]
6. Vincent, A.T.; Schiettekatte, O.; Goarant, C.; Neela, V.K.; Bernet, E.; Thibeaux, R.; Ismail, N.; Khalid, M.K.N.M.; Amran, F.; Masuzawa, T.; et al. Revisiting the taxonomy and evolution of pathogenicity of the genus Leptospira through the prism of genomics. *PLoS Negl. Trop. Dis.* **2019**, *13*, e0007270. [CrossRef]
7. Guglielmini, J.; Bourhy, P.; Schiettekatte, O.; Zinini, F.; Brisse, S.; Picardeau, M. Genus-wide Leptospira core genome multilocus sequence typing for strain taxonomy and global surveillance. *PLoS Negl. Trop. Dis.* **2019**, *13*, e0007374. [CrossRef]
8. Barragan, V.; Chiriboga, J.; Miller, E.; Olivas, S.; Birdsell, D.; Hepp, C.; Hornstra, H.; Schupp, J.M.; Morales, M.; Gonzalez, M.; et al. High Leptospira Diversity in Animals and Humans Complicates the Search for Common Reservoirs of Human Disease in Rural Ecuador. *PLoS Negl. Trop. Dis.* **2016**, *10*. [CrossRef]
9. Bett, B.; Kiunga, P.; Gachohi, J.; Sindato, C.; Mbotha, D.; Robinson, T.; Lindahl, J.; Grace, D. Effects of climate change on the occurrence and distribution of livestock diseases. *Prev. Vet. Med.* **2017**, *137*, 119–129. [CrossRef]
10. Guerra, M.A. Leptospirosis: Public health perspectives. *Biologicals* **2013**. [CrossRef]
11. Guernier, V.; Goarant, C.; Benschop, J.; Lau, C.L. A systematic review of human and animal leptospirosis in the Pacific Islands reveals pathogen and reservoir diversity. *PLoS Negl. Trop. Dis.* **2018**, *12*. [CrossRef]

12. Guernier, V.; Richard, V.; Nhan, T.; Rouault, E.; Tessier, A.; Musso, D. Leptospira diversity in animals and humans in Tahiti, French Polynesia. *PLoS Negl. Trop. Dis.* **2017**, *11*. [CrossRef] [PubMed]
13. Markovych, O.; Tymchyk, V.; Kolesnikova, I. Leptospirosis in Zakarpattia Oblast (2005–2015). *Vector-Borne Zoonotic Dis.* **2019**, *19*, 333–340. [CrossRef] [PubMed]
14. Dhewantara, P.W.; Hu, W.; Zhang, W.; Yin, W.; Ding, F.; Mamun, A.; Soares Magalhaes, R.J. Leptospirosis, Climate and Satellite-based Environmental Factors: A Temporal Modeling. *Online J. Public Health Inform.* **2019**, *11*, 3–4. [CrossRef]
15. Ceccato, P.; Ramirez, B.; Manyangadze, T.; Gwakisa, P.; Thomson, M.C. Data and tools to integrate climate and environmental information into public health. *Infect. Dis. Poverty* **2018**, *7*, 126. [CrossRef] [PubMed]
16. Lau, C.L.; Smythe, L.D.; Craig, S.B.; Weinstein, P. Climate change, flooding, urbanisation and leptospirosis: Fuelling the fire? *Trans. R. Soc. Trop. Med. Hyg.* **2010**, *104*, 631–638. [CrossRef] [PubMed]
17. Schneider, A.G.; Casanovas-Massana, A.; Hacker, K.P.; Wunder, E.A.; Begon, M.; Reis, M.G.; Childs, J.E.; Costa, F.; Lindow, J.C.; Ko, A.I. Quantification of pathogenic Leptospira in the soils of a Brazilian urban slum. *PLoS Negl. Trop. Dis.* **2018**, *12*. [CrossRef]
18. Thibeaux, R.; Geroult, S.; Benezech, C.; Chabaud, S.; Soupé-Gilbert, M.E.; Girault, D.; Bierque, E.; Goarant, C. Seeking the environmental source of Leptospirosis reveals durable bacterial viability in river soils. *PLoS Negl. Trop. Dis.* **2017**, *11*. [CrossRef]
19. Henry, R.A.; Johnson, R.C. Distribution of the genus Leptospira in soil and water. *Appl. Environ. Microbiol.* **1978**, *35*, 492–499. [CrossRef]
20. Baker, M.F.; Baker, H.J. Pathogenic Leptospira in Malaysian surface waters. I. A method of survey for Leptospira in natural waters and soils. *Am. J. Trop. Med. Hyg.* **1970**, *19*, 485–492. [CrossRef]
21. Rood, E.J.J.; Goris, M.G.A.; Pijnacker, R.; Bakker, M.I.; Hartskeerl, R.A. Environmental risk of leptospirosis infections in the Netherlands: Spatial modelling of environmental risk factors of leptospirosis in the Netherlands. *PLoS ONE* **2017**, *12*, e0186987. [CrossRef]
22. Gracie, R.; Barcellos, C.; Magalhães, M.; Souza-Santos, R.; Guimarães Barrocas, P.R. Geographical scale effects on the analysis of leptospirosis determinants. *Int. J. Environ. Res. Public Health* **2014**, *11*, 10366–10383. [CrossRef] [PubMed]
23. Vieira, A.S.; Pinto, P.S.; Lilenbaum, W. A systematic review of leptospirosis on wild animals in Latin America. *Trop. Anim. Health Prod.* **2018**, *50*, 229–238. [CrossRef] [PubMed]
24. Sanhueza, J.M.; Heuer, C.; Wilson, P.R.; Benschop, J.; Collins-Emerson, J.M. Prevalence and risk factors for *Leptospira* exposure in New Zealand veterinarians. *Epidemiol. Infect.* **2015**, *143*, 2116–2125. [CrossRef] [PubMed]
25. Lau, C.L.; Watson, C.H.; Lowry, J.H.; David, M.C.; Craig, S.B.; Wynwood, S.J.; Kama, M.; Nilles, E.J. Human Leptospirosis Infection in Fiji: An Eco-epidemiological Approach to Identifying Risk Factors and Environmental Drivers for Transmission. *PLoS Negl. Trop. Dis.* **2016**. [CrossRef] [PubMed]
26. Le Turnier, P.; Epelboin, L. Update on leptospirosis. *Rev. Med. Interne* **2019**, *40*, 306–312. [CrossRef]
27. López, M.S.; Müller, G.V.; Lovino, M.A.; Gómez, A.A.; Sione, W.F.; Aragonés Pomares, L. Spatio-temporal analysis of leptospirosis incidence and its relationship with hydroclimatic indicators in northeastern Argentina. *Sci. Total Environ.* **2019**, *694*. [CrossRef]
28. Mwachui, M.A.; Crump, L.; Hartskeerl, R.; Zinsstag, J.; Hattendorf, J. Environmental and Behavioural Determinants of Leptospirosis Transmission: A Systematic Review. *PLoS Negl. Trop. Dis.* **2015**, *9*, e0003843. [CrossRef]
29. Suwanpakdee, S.; Kaewkungwal, J.; White, L.J.; Asensio, N.; Ratanakorn, P.; Singhasivanon, P.; Day, N.P.J.; Pan-Ngum, W. Spatio-temporal patterns of leptospirosis in Thailand: Is flooding a risk factor? *Epidemiol. Infect.* **2015**. [CrossRef]
30. Ngugi, J.N.; Fèvre, E.M.; Mgode, G.F.; Obonyo, M.; Mhamphi, G.G.; Otieno, C.A.; Cook, E.A.J. Seroprevalence and associated risk factors of leptospirosis in slaughter pigs; A neglected public health risk, western Kenya. *BMC Vet. Res.* **2019**, *15*, 1–11. [CrossRef]
31. Pinto, P.S.; Libonati, H.; Lilenbaum, W. A systematic review of leptospirosis on dogs, pigs, and horses in Latin America. *Trop. Anim. Health Prod.* **2017**, *49*. [CrossRef]
32. Yatbantoong, N.; Chaiyarat, R. Factors associated with leptospirosis in domestic cattle in salakphra wildlife sanctuary, Thailand. *Int. J. Environ. Res. Public Health* **2019**, *16*, 1042. [CrossRef] [PubMed]

33. Cosson, J.-F.; Picardeau, M.; Mielcarek, M.; Tatard, C.; Chaval, Y.; Suputtamongkol, Y.; Buchy, P.; Jittapalapong, S.; Herbreteau, V.; Morand, S. Epidemiology of Leptospira Transmitted by Rodents in Southeast Asia. *PLoS Negl. Trop. Dis.* **2014**, *8*, e2902. [CrossRef] [PubMed]
34. Weis, S.; Hartmann, K. Infektionen mit Leptospiren bei der Katze. *Tierarztl. Prax. Ausg. K Kleintiere-Heimtiere* **2017**, *45*, 103–108. [CrossRef] [PubMed]
35. Sant'Anna, R.; Vieira, A.S.; Oliveira, J.; Lilenbaum, W. Asymptomatic leptospiral infection is associated with canine chronic kidney disease. *Comp. Immunol. Microbiol. Infect. Dis.* **2019**, *62*, 64–67. [CrossRef]
36. Weinberger, D.; Baroux, N.; Grangeon, J.P.; Ko, A.I.; Goarant, C. El Niño Southern Oscillation and Leptospirosis Outbreaks in New Caledonia. *PLoS Negl. Trop. Dis.* **2014**, *8*, e2798. [CrossRef]
37. Dufour, B.; Moutou, F.; Hattenberger, A.M.; Rodhain, F. Global change: Impact, management, risk approach and health measures—The case of Europe. *Rev. Sci. Tech.* **2008**, *27*, 529–550. [CrossRef]
38. Martins, G.; Lilenbaum, W. Control of bovine leptospirosis: Aspects for consideration in a tropical environment. *Res. Vet. Sci.* **2017**, *112*, 156–160. [CrossRef]
39. Dhewantara, P.W.; Lau, C.L.; Allan, K.J.; Hu, W.; Zhang, W.; Mamun, A.A.; Soares Magalhães, R.J. Spatial epidemiological approaches to inform leptospirosis surveillance and control: A systematic review and critical appraisal of methods. *Zoonoses Public Health* **2019**, *66*, 185–206. [CrossRef]
40. Zhao, J.; Liao, J.; Huang, X.; Zhao, J.; Wang, Y.; Ren, J.; Wang, X.; Ding, F. Mapping risk of leptospirosis in China using environmental and socioeconomic data. *BMC Infect. Dis.* **2016**, *16*, 343. [CrossRef]
41. Jara, M.; Escobar, L.E.; Rodriges, R.O.; Frias-De-Diego, A.; Sanhueza, J.; Machado, G. Spatial distribution and spread potential of sixteen Leptospira serovars in a subtropical region of Brazil. *Transbound. Emerg. Dis.* **2019**, *66*, 2482–2495. [CrossRef]
42. Thibeaux, R.; Iraola, G.; Ferrés, I.; Bierque, E.; Girault, D.; Soupé-Gilbert, M.E.; Picardeau, M.; Goarant, C. Deciphering the unexplored Leptospira diversity from soils uncovers genomic evolution to virulence. *Microb. Genom.* **2018**, *4*. [CrossRef]
43. Casanovas-Massana, A.; Pedra, G.G.; Wunder, E.A.; Diggle, P.J.; Begon, M.; Ko, A.I. Quantification of Leptospira interrogans survival in soil and water microcosms. *Appl. Environ. Microbiol.* **2018**, *84*. [CrossRef] [PubMed]
44. Wilson, R.; Fujioka, R. Development of a method to selectively isolate pathogenic Leptospira from environmental samples. *Water Sci. Technol.* **1995**, *31*, 275–282. [CrossRef]
45. Diesch, S.L.; McCulloch, W.F. Isolation of pathogenic leptospires from waters used for recreation. *Public Health Rep.* **1966**, *81*, 299–304. [CrossRef] [PubMed]
46. Ismail, S.; Wahab, N.Z.A.; Badya, N.; Rahman, N.I.A.; Yeo, C.C.; Latif, A.Z.A.; Haque, M. A study on the presence of pathogenic leptospira spp. in environmental water samples obtained from selected recreational areas in terengganu, malaysia. *Res. J. Pharm. Technol.* **2014**, *7*, 1153–1157.
47. Lall, C.; Kumar, K.V.; Raj, R.V.; Vedhagiri, K.; Sunish, I.P.; Vijayachari, P. Correlation between physicochemical properties of soil and presence of leptospira. *Ecohealth* **2018**, *15*, 670–675. [CrossRef]
48. Wójcik-Fatla, A.; Zajac, V.; Wasinski, B.; Sroka, J.; Cisak, E.; Sawczyn, A.; Dutkiewicz, J. Occurrence of Leptospira DNA in water and soil samples collected in eastern Poland. *Ann. Agric. Environ. Med.* **2014**, *21*, 730–732. [CrossRef]
49. Phillips, S.J.; Anderson, R.P.; Schapire, R.E. Maximum entropy modeling of species geographic distributions. *Ecol. Model.* **2006**, *190*, 231–259. [CrossRef]
50. Elith, J.; Phillips, S.J.; Hastie, T.; Dudík, M.; Chee, Y.E.; Yates, C.J. A statistical explanation of MaxEnt for ecologists. *Divers. Distrib.* **2011**, *17*, 43–57. [CrossRef]
51. Franklin, J.; Miller, J.A. *Mapping Species Distributions: Spatial Inference and Prediction*; Cambridge University Press: Cambridge, UK, 2010; ISBN 9780511810602. [CrossRef]
52. Peterson, A.T.; Papeş, M.; Soberón, J. Rethinking receiver operating characteristic analysis applications in ecological niche modeling. *Ecol. Model.* **2008**. [CrossRef]
53. Abdrakhmanov, S.K.; Mukhanbetkaliyev, Y.Y.; Korennoy, F.I.; Sultanov, A.A.; Kadyrov, A.S.; Kushubaev, D.B.; Bakishev, T.G. Maximum entropy modeling risk of anthrax in the Republic of Kazakhstan. *Prev. Vet. Med.* **2017**, *144*, 149–157. [CrossRef]
54. Abdrakhmanov, S.K.; Sultanov, A.A.; Beisembayev, K.K.; Korennoy, F.I.; Kushubaev, D.B.; Kadyrov, A.S. Zoning the territory of the Republic of Kazakhstan as to the risk of rabies among various categories of animals. *Geospat. Health* **2016**, *11*, 174–181. [CrossRef] [PubMed]

55. Volodin, E.M.; Dianskii, N.A.; Gusev, A.V. Simulating present-day climate with the INMCM4.0 coupled model of the atmospheric and oceanic general circulations. *Izv.—Atmos. Ocean Phys.* **2010**, *46*, 414–431. [CrossRef]
56. Andre-Fontaine, G.; Aviat, F.; Thorin, C. Waterborne Leptospirosis: Survival and Preservation of the Virulence of Pathogenic Leptospira spp. in Fresh Water. *Curr. Microbiol.* **2015**, *71*, 136–142. [CrossRef]
57. Egorov, V.A.; Bartalev, S.A.; Kolbudaev, P.A.; Plotnikov, D.E.; Khvostikov, S.A. Land cover map of Russia derived from Proba-V satellite data. *Sovrem. Probl. Distantsionnogo Zo. Zemli iz Kosmosa* **2018**, *15*, 282–286. [CrossRef]
58. Broxton, P.D.; Zeng, X.; Scheftic, W.; Troch, P.A. A MODIS-Based Global 1-km Maximum Green Vegetation Fraction Dataset. *J. Appl. Meteorol. Climatol.* **2014**, *53*, 1996–2004. [CrossRef]
59. Hengl, T.; De Jesus, J.M.; Heuvelink, G.B.M.; Gonzalez, M.R.; Kilibarda, M.; Blagotić, A.; Shangguan, W.; Wright, M.N.; Geng, X.; Bauer-Marschallinger, B.; et al. SoilGrids250m: Global gridded soil information based on machine learning. *PLoS ONE* **2017**. [CrossRef]
60. Naimi, B.; Hamm, N.A.S.; Groen, T.A.; Skidmore, A.K.; Toxopeus, A.G. Where is positional uncertainty a problem for species distribution modelling? *Ecography (Cop.)* **2014**, *37*, 191–203. [CrossRef]
61. Phillips, S.J.; Dudík, M. Modeling of species distributions with Maxent: New extensions and a comprehensive evaluation. *Ecography (Cop.)* **2008**, *31*, 161–175. [CrossRef]

© 2020 by the authors. Licensee MDPI, Basel, Switzerland. This article is an open access article distributed under the terms and conditions of the Creative Commons Attribution (CC BY) license (http://creativecommons.org/licenses/by/4.0/).

pathogens

MDPI

Article

Leptospira Survey in Wild Boar (*Sus scrofa*) Hunted in Tuscany, Central Italy

Giovanni Cilia, Fabrizio Bertelloni *, Marta Angelini, Domenico Cerri and Filippo Fratini

Department of Veterinary Sciences, University of Pisa, Viale delle Piagge 2, 56124 Pisa, Italy; giovanni.cilia@vet.unipi.it (G.C.); m.angelini5@studenti.unipi.it (M.A.); domenico.cerri@unipi.it (D.C.); filippo.fratini@unipi.it (F.F.)

* Correspondence: fabrizio.bertelloni@unipi.it; Tel.: +39-050-2216-969

Received: 20 April 2020; Accepted: 12 May 2020; Published: 14 May 2020

check for updates

Abstract: Leptospirosis is a re-emerging, worldwide zoonosis, and wild boar (*Sus scrofa*) are involved in its epidemiology as the reservoir. The aim of this study was to investigate the prevalence of *Leptospira* with serological, bacteriological, and molecular assays in wild boar hunted in Tuscany (Italy) during two hunting seasons. In total, 287 specimens of sera, kidneys, and liver were collected to perform microscopic agglutination tests (MATs), isolation, and RealTime PCR to detect pathogenic (*lipL32* gene), intermediate (*16S rRNA* gene), and saprophytic (*23S rRNA* gene) *Leptospira*. Within sera, 39 (13.59%) were positive to the MAT, and Australis was the most represented serogroup (4.88%), followed by Pomona (4.18%), and Tarassovi (3.14%). Moreover, four *Leptospira* cultures were positive, and once isolates were identified, one was identified as *L. borgpetersenii* serovar Tarassovi, and three as *L. interrogans* serovar Bratislava. Pathogenic *Leptospira* DNA were detected in 32 wild boar kidneys (11.15%). The characterization through the amplification of the *rrs2* gene highlighted their belonging to *L. interrogans* (23 kidneys), *L. borgpetersenii* (four), and *L. kirschneri* (one), while nine kidneys (3.14%) were positive for intermediate *Leptospira*, all belonging to *L. fainei*. The results of this study confirmed the importance of wild boar in the epidemiology of leptospirosis among wildlife in Central Italy.

Keywords: leptospirosis; zoonosis; infectious disease; multilocus sequence typing (MLST); wildlife; *Leptospira fainei*; MAT; intermediate *Leptospira*

1. Introduction

Wild boar (*Sus scrofa*) is a large ungulate mammal with worldwide distribution. It can live in several types of habitat, including urban and suburban areas [1,2]. Due to their high adaptability, wild boar populations have rapidly increased in number during recent years, in Europe, and especially in Italy [1,3]. In Italy, wild boar is largely spread in all areas, from the Alps to the southern part of the Italian peninsula, including the islands. There is a high density, particularly in specific regions, such as Tuscany [3–5]. The abundant presence of wild boar in the Tuscany region, as well in Central Italy, is suggested by the very high number of animals hunted in this area; every year the hunting of about 42,000 specimens is registered [1,3–5]. The massive presence of wild boar in particular areas, other than representing an important source of damage for agriculture [6], can be a severe risk to human and animal health, due to the identification of wild boar as reservoir for many etiological agents; among them typical zoonoses, such as *Leptospira* [7–9].

Leptospirosis is a re-emerging zoonotic disease with worldwide spread. It is caused by *Leptospira* spp., a Gram-negative spirochetal bacterium [10–12]. The genus *Leptospira* is divided into more than 260 antigenically-different serovars, classified as pathogenic, intermediate, and saprophytic, with different levels of pathogenicity for animals and humans [13,14]. While pathogenic *Leptospira* cause mild or severe infection, intermediate *Leptospira* could possibly be pathogenic, causing mild infection, while saprophytic *Leptospira* are present in the environment and are non-pathogenic [13,14]. Intermediate and saprophytic *Leptospira* could be important due to the strictly-contact and recombination events with pathogenic serovars [15–17]. Leptospirosis occurs in tropical, subtropical, and temperate zones, where it is maintained by a large variety of both wild and domestic mammals which can play the role of *Leptospira* maintenance host [18–21]. The reservoir organisms generally do not develop symptoms, except after a long time [11,12].

Leptospira renal-carrying/-colonization/-localization in asymptomatic animals contributes to the maintenance of infection in a particular environment by constantly shedding bacteria through their urine. Accidental contact with *Leptospira*-infected animal urine causes incidental infection, and produces clinical diseases in so-called "incidental hosts" [11,21].

Swine, including wild boar and pig, are recognized as maintenance hosts for Pomona, Tarassovi, and Bratislava serovars [21], but can be infected by other *Leptospira* serovars, in relation to both geographic area where the population lives and their behavior [22–26]. The epidemiology of leptospirosis may change over time in domestic and wild animals, and some serovars seems to be prevalent and emerging [26,27]. Moreover, intermediate *Leptospira* DNA has been detected in the kidneys of wild boar hunted in Liguria region (Italy), suggesting a possible infection [7].

Tuscany, as well as all of Central Italy, is a geographic area that promotes the presence and the persistence of *Leptospira* in the ecosystem. The features of *Leptospira*-spreading are the presence of several wild animals involved as reservoir, domestic animals bred in extensive farms in contact with wildlife, high presence of hunting activity, and abundance of wetlands, such as marshes, ponds, and irrigation canals [9,26,28–33].

The aim of this investigation was to detect and characterize pathogenic, intermediate, and saprophytic *Leptospira* in wild boar hunted in Tuscany region during two hunting seasons (2018/ 2019 and 2019/2020), in order to delineate the risk for the transmission and spreading of leptospirosis to domestic animals and humans.

2. Results

Serum, kidney, and liver samples were collected from a total of 287 hunted wild boar. Two hundred wild boar were sampled during 2018/2019 hunting season— 75 from Grosseto province, 58 from Pisa province, 55 from Siena province, and 12 from Livorno province (Figure 1). In addition, 87 specimens were sampled during 2019/2020 hunting seasons with 38, 37, and 12 from Pisa, Grosseto, and Lucca provinces, respectively (Figure 1).

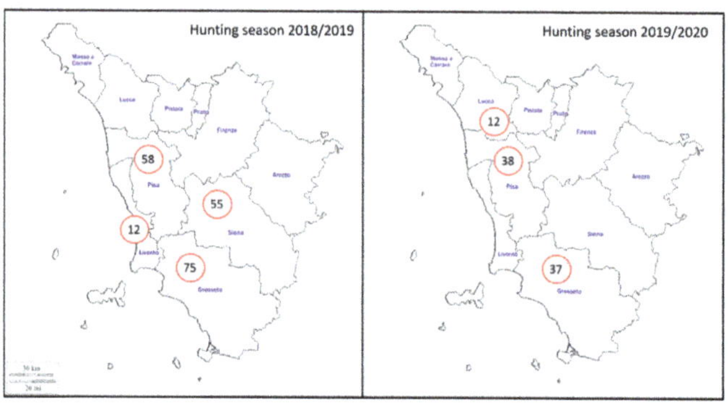

A B

Figure 1. Geographical distribution of the sampling area included in the study (Tuscany region, Italy). The number of sampled hunted wild boar per province is indicated in relation to hunting seasons. (**A**) Hunting season 2018/2019; (**B**) Hunting season 2019/2020.

Results on distribution of positive sera and kidney for pathogenic *Leptospira* in relation to hunting season, province, sex, and age class are reported in Table 1.

Table 1. Distribution of positive sera and kidney for pathogenic *Leptospira* in relation to hunting season, province, sex, and age class.

Hunting Season	Province	Sex	Age Class	Examined Wild Boar	MAT-Positive Sera (%)	PCR-Positive Kidneys (%)
2018/2019 (n = 200)	Pisa (n = 58)	Male (n = 30)	Adult	9	2 (22.2)	1 (11.1)
			Subadult	10	2 (20.0)	4 (40.0)
			Young	11	3 (27.3)	0
		Female (n = 28)	Adult	14	2 (14.3)	2 (14.3)
			Subadult	5	1 (20.0)	2 (40.0)
			Young	9	1 (11.1)	2 (22.2)
	Grosseto (n = 75)	Male (n = 29)	Adult	10	2 (20.0)	2 (20.0)
			Subadult	5	1 (20.0)	0
			Young	14	1 (7.1)	3 (21.4)
		Female (n = 46)	Adult	22	2 (9.09)	1 (4.6)
			Subadult	5	0	0
			Young	19	2 (10.5)	4 (21.1)
	Siena (n = 55)	Male (n = 22)	Adult	10	2 (20.0)	0
			Subadult	4	1 (25.0)	0
			Young	8	0	1 (12.5)
		Female (n = 33)	Adult	21	5 (23.8)	3 (14.3)
			Subadult	2	0	0
			Young	10	1 (10.0)	3 (30.0)
	Livorno (n = 12)	Male (n = 4)	Adult	2	0	0
			Subadult	0	0	0
			Young	2	1 (50.0)	1 (50.0)
		Female (n=8)	Adult	4	1 (25.0)	1 (25.0)
			Subadult	0	0	0
			Young	4	2 (50.0)	1 (25.0)
2019/2020 (n = 87)	Pisa (n = 38)	Male (n = 13)	Adult	6	0	0
			Subadult	4	0	0
			Young	3	0	0
		Female (n = 25)	Adult	21	2 (9.52)	0
			Subadult	1	1 (100)	0
			Young	3	0	1 (33.3)
	Grosseto (n = 37)	Male (n = 16)	Adult	11	1 (9.09)	0
			Subadult	1	0	0
			Young	4	0	0
		Female (n = 21)	Adult	10	1 (10.0)	0
			Subadult	5	1 (20.0)	0
			Young	6	1 (16.7)	0
	Lucca (n=12)	Male (n = 4)	Adult	1	0	0
			Subadult	0	0	0
			Young	3	0	0
		Female (n = 8)	Adult	4	0	0
			Subadult	0	0	0
			Young	4	0	0

2.1. Microscopic Agglutination Test (MAT)

Overall, 39 out of 287 sera (13.59%) were positive in the MAT (Table 2). Considering each hunting season, seropositivity of 16% (32 out of 200 sera) was recorded during 2018/2019, while 8.05% (7 out of 87) was recorded during 2019/2020. Considering wild boar sex, 16 out of 118 male sera (13.55%) and 23 out of 169 (13.61%) were positive in the MAT. Moreover, in relation to age class, 20 out of 142 adult specimens' sera (14.08%), 7 out of 42 subadult specimens' sera (16.67%) and 12 out of 100 young specimens' sera (12.00%) were positive in serological analysis.

Table 2. Numbers of positive serological reactions detected for wild boar sera in relation to different *Leptospira* serogroups at low (1:100) and high titers (1:12,800).

Leptospira Serogroup	Titer								Total (%)
	100	200	400	800	1600	3200	6400	12800	
Icterohaemorrhagiae			1						1 (2.56%)
Canicola		1	1						2 (5.13%)
Pomona	8	1	3						12 (30.8%)
Grippotyphosa									
Tarassovi	4	1			1	1	1	1	9 (23.1%)
Australis	5		5	2	1	1			14 (35.9%)
Sejroe									
Ballum							1		1 (2.56%)
Total	17	3	10	2	2	2	2	1	39 (100%)

Australis resulted the most-recorded serogroup (4.88%), followed by Pomona (4.18%), and Tarassovi (3.14%). Other antibody titers were reported for serogroup Canicola (0.70%) and for serogroups Icterohaemorrhagiae and Ballum (0.45%). The highest titer detected was 1:12,800 for serogroup Tarassovi, followed by titer of 1:6400, which was reported for serogroups Tarassovi and Ballum.

Results on distribution of positive sera detected by MAT in relation to hunting season, province, sex, and age class are reported in Table 1. No statistical differences ($p > 0.05$) were reported for the serological positivity considering hunting seasons, provinces, and wild boar sex and age class. Moreover, comparing all parameters, no statistical differences ($p > 0.05$) were showed in Pisa and Grosseto during the two different hunting seasons.

2.2. Molecular Analysis

Concerning pathogenic *Leptospira,* DNA was detected in 11.15% (32 out of 297) of wild boar kidneys. Table 1 shows PCR-positive kidneys in relation to hunting seasons, province and wild boar sex and age class. During the 2018/2019 and 2019/2020 hunting seasons, 15.5% (31 out of 200) and 1.15% (1 out of 87) of PCR positivity was reported among kidneys samples, respectively. Considering wild boar sex, 12 out of 118 male sera (10.16%) and 20 out of 169 (11.83%) scored positive. Moreover, in relation to age class, 10 out of 142 adult specimens' kidneys (7.04%), 6 out of 42 subadult specimens' kidneys (14.28%), and 16 out of 100 young specimens' kidneys (16.00%) gave positive results in serological analysis.

No statistical differences ($p > 0.05$) were highlighted comparing province, wild boar sex, or age class. Conversely, the incidence of pathogenic *Leptospira*-positive kidneys was statically higher ($p \leq 0.01$) during 2018/2019 hunting season compared to the 2019/2020 ones.

The detection of pathogenic *Leptospira* DNA was higher ($p \leq 0.01$) during 2018/2019 hunting season in both Pisa and Grosseto provinces compared to the second hunting season. On the contrary, there were no statistical differences ($p > 0.05$) in the Pisa and Grosseto provinces during the two different hunting seasons, comparing sex and age class of wild boar.

The 3.14% (9 out of 287) of kidneys were positive for intermediate *Leptospira*. The positivity in relation to hunting seasons, province, wild boar sex, and age class are showed in Table 1. All the intermediate *Leptospira*-positive kidneys (4.5%; 9 out of 200) were collected in 2018/2019, highlighting a statistical difference ($p \leq 0.01$) in relation to 2019/2020 hunting season. Also, the results showed a statistically-higher infection rate ($p \leq 0.01$) in male compared to female, and in Pisa province compared to other provinces. No statistical difference ($p > 0.05$) were noted among age classes.

No saprophytic *Leptospira* DNA was detected in kidney samples. No positive reactions were recorded in wild boar livers across all specimens during the two year of investigation.

2.3. *Leptospira spp. Isolation, Characterization and Genotyping*

Four *Leptospira* cultures were positive after 30 days of incubation. The results, reported in Table 3, show that three isolates were obtained from subadult males hunted in Pisa province, while the other one was from an adult female hunted in Livorno. Through multilocus sequence typing (MLST) analysis, one isolate was identified as *Leptospira borgpetersenii* serogroup Tarassovi serovar Tarassovi (Sequence Type 153), while the other three were identified as *L. interrogans* serogroup Bratislava serovar Bratislava (ST 24), as reported in Table 3. Moreover, the amplification of the *rrs2* gene from kidney tissue highlights that the species belonged to *L. borgpetersenii* and *L. interrogans*, respectively.

Table 3. Characterization of wild boar *Leptospira* isolates tested with anti-sera and multilocus sequence typing (MLST).

Sample	Wild Boar			Isolates Characterization	
	Sex	Age Class	Province	Anti-Serum MAT Serogroup	MLST (Sequence Type)
Kidney 5	Male	Subadult	Pisa	Tarassovi	Tarassovi (ST 153)
Kidney 14	Male	Subadult	Pisa	Australis	Bratislava (ST 24)
Kidney 15	Male	Subadult	Pisa	Australis	Bratislava (ST 24)
Kidney 22	Female	Adult	Livorno	Australis	Bratislava (ST 24)

With regard to characterization of PCR-positive samples, amplification of the *rr2* gene highlighted that pathogenic *Leptospira* belonged to *L. interrogans* (23 kidneys), *L. borgpetersenii* (four) and *L. kirschneri* (one) (Table 4). Moreover, phylogenetic analysis identified the close relationship to their respective *Leptospira* species. (Figure 2).

Table 4. Characterization of *Leptospira* species in wild boar pathogenic *Leptospira*-positive PCR-amplifying *rrs2* gene.

Sample	Wild Boar			Isolate Characterization
	Sex	Age Class	Province	*Leptospira* Species
Kidney 1	Female	Young	Pisa	*L. borgpetersenii*
Kidney 2	Female	Subadult	Pisa	*L. interrogans*
Kidney 3	Male	Adult	Pisa	*L. borgpetersenii*
Kidney 4	Female	Young	Pisa	*L. interrogans*
Kidney 6	Male	Young	Siena	*L. interrogans*
Kidney 7	Female	Young	Siena	*L. interrogans*
Kidney 12	Female	Young	Siena	*L. interrogans*
Kidney13	Female	Adult	Siena	*L. interrogans*
Kidney 20	Male	Young	Grosseto	*L. interrogans*
Kidney 24	Female	Young	Livorno	*L. interrogans*
Kidney 36	Female	Adult	Grosseto	*L. interrogans*
Kidney39	Female	Young	Grosseto	*L. interrogans*
Kidney 42	Female	Young	Siena	*L. interrogans*
Kidney 53	Female	Young	Grosseto	*L. interrogans*
Kidney 54	Male	Young	Grosseto	*L. kirschneri*
Kidney 55	Male	Young	Grosseto	*L. interrogans*
Kidney 58	Female	Adult	Pisa	*L. borgpetersenii*
Kidney 64	Female	Adult	Siena	*L. interrogans*
Kidney 65	Female	Adult	Siena	*L. interrogans*
Kidney 67	Female	Adult	Pisa	*L. interrogans*
Kidney 68	Female	Subadult	Pisa	*L. interrogans*
Kidney 71	Male	Adult	Grosseto	*L. interrogans*
Kidney 72	Female	Young	Grosseto	*L. borgpetersenii*
Kidney 75	Male	Subadult	Pisa	*L. interrogans*
Kidney 78	Male	Young	Livorno	*L. interrogans*
Kidney 97	Male	Adult	Grosseto	*L. interrogans*
Kidney 165	Female	Young	Grosseto	*L. interrogans*
Kidney 208	Female	Young	Pisa	*L. interrogans*

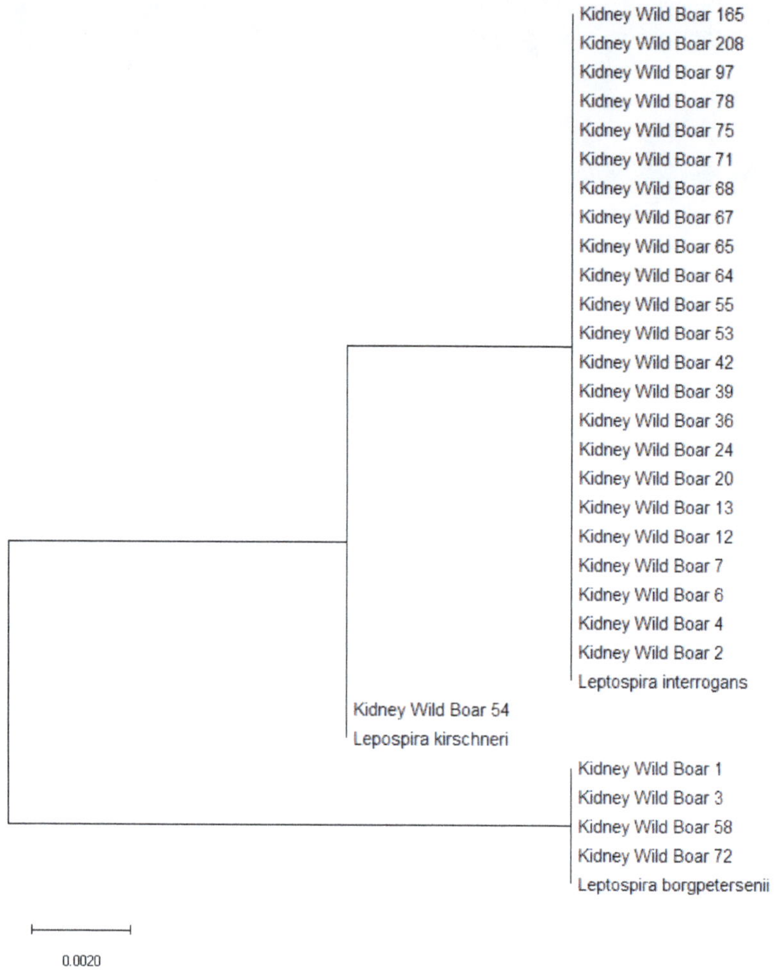

Figure 2. Molecular phylogenetic analysis for the *rrs2* gene of *Leptospira interrogans, Leptospira borgpetersenii,* and *Leptospira kirschneri* by the maximum likelihood method, based on the Tamura–Nei model. The branch lengths of the tree measured the number of substitutions per site. The analysis involved 31 nucleotide sequences. There was a total of 452 positions in the final dataset.

Moreover, the amplification of intermediate *Leptospira 16s rRNA* DNA of PCR-positive specimens showed *L. fainei in* all nine kidneys (Table 5). Furthermore, the phylogenetic analysis identified the close relationship to *L. fainei* specie. (Figure 3).

Table 5. Characterization of *Leptospira* species in wild boar intermediate *Leptospira*-positive PCR-amplifying *16s rRNA* gene.

Sample	Wild Boar			Isolate Characterization
	Sex	Age Class	Province	*Leptospira* Species
Kidney 23	Male	Young	Livorno	*L. fainei*
Kidney 27	Male	Adult	Pisa	*L. fainei*
Kidney 31	Female	Adult	Pisa	*L. fainei*
Kidney 56	Male	Young	Grosseto	*L. fainei*
Kidney 57	Male	Adult	Pisa	*L. fainei*
Kidney 63	Male	Adult	Siena	*L. fainei*
Kidney 69	Female	Subadult	Pisa	*L. fainei*
Kidney123	Male	Adult	Livorno	*L. fainei*
Kidney 153	Male	Adult	Siena	*L. fainei*

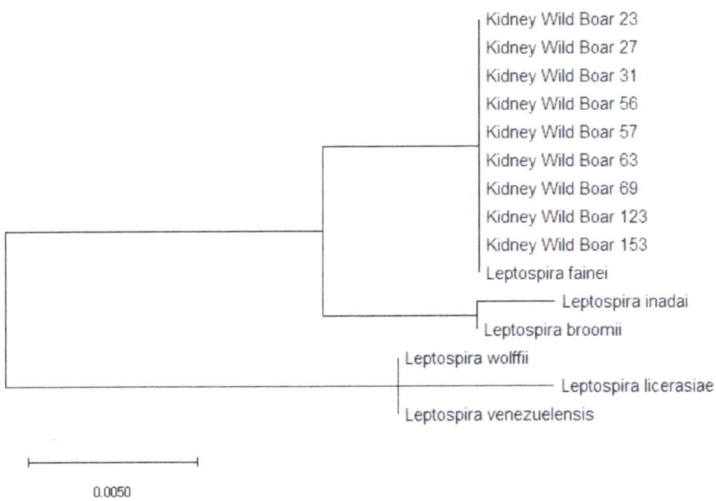

Figure 3. Molecular phylogenetic analysis for 16s rRNA gene of *Leptospira fainei, Leptospira inadai, Leptospira broomii, Leptospira wolffii, Leptospira licerasiae,* and *Leptospira venezuelensis* by the maximum likelihood method based on the Tamura–Nei model. The branch lengths of the tree measured the number of substitutions per site. The analysis involved 22 nucleotide sequences. There was a total of 438 positions in the final dataset.

3. Discussion

Leptospirosis is a re-emerging worldwide public health risk, but is underestimated and characterized by a downward trend [34]. Climatic changes, rainfall, modifications of ecological niches, and new potential maintenance hosts all represent important features involved in *Leptospira* epidemiology.

Wild boar, among wildlife, is an important *Leptospira* reservoir and, for several areas, represents an appropriate indicator for this zoonotic infectious disease.

In this investigation we reported the results of serological analysis, isolation and molecular investigations performed on 287 hunted wild boar during two hunting seasons (2018/2019 and 2019/2020).

With regard to serological assay, the prevalence of *Leptospira* infection, recorded in both hunting seasons, was very similar to other studies carried out on wild boar in Tuscany [9,22,26,31,33]. Moreover, the seroprevalence reported in this investigation was very close to other data obtained in different

Italian regions [27,35–38]. Unfortunately, serological data about leptospirosis in wildlife, especially regarding wild boar, are available just in some regions. It also seems that *Leptospira* serovars/serogroups have a different geographical distribution, suggesting a distinct circulation/epidemiology in other environments/ecosystems. Examining each region, Australis, Pomona, and Tarassovi, are the most-detected serogroups in the Tuscany region [22,26,31,33], In the Lombardy and Emilia Romagna regions it is Bratislava [37–39], in the Campania region it is Tarassovi [35], whereas in the Sardinia region it is Pomona and Grippotyphosa [36].

The distribution of *Leptospira* serovars in wild boar in Europe is also not homogeneous; high levels of Pomona infection was recorded in Germany, Croatia, Poland, and Spain [23,40–42]. Bratislava was the most-detected serovar in Sweden [25], Tarassovi in Portugal and Slovenia [24,43], Grippotyphosa in Czech Republic [44], and Hardjo in Poland [23].

Little information is available on *Leptospira* isolation in wild boar, especially in Italy [38]. The obtained *Leptospira* isolates, identified by MLST, confirm the circulation of Tarassovi and Bratislava serogroups within wild boar in Tuscany. Bratislava isolation is commonly performed in wildlife due to the high spectrum of maintenance hosts [45–47], while Tarassovi is rarely isolated and detected through serology. Indeed, Tarassovi is strictly a swine-specific serovar; its isolation, reported in this investigation, seems to confirm the hypothesis that wild boar could serve as reservoir of Tarassovi [26,43,48]. Only two of them (Bratislava, isolated from subadult from Pisa province; Table 3) reported correlated serological positivity for serogroup Australis at titer 1:100, while the other two gave negative results in the MAT. No correlation was found between the MAT and PCR-positive results. The seronegativity of *Leptospira*-positive kidneys has been previously reported for other animal species [49–52], including swine [47], suggesting an early or chronic infection.

Conversely to serological results, very few studies were performed on pathogenic *Leptospira* DNA in wild boar kidneys. In spite of this, the prevalence of pathogenic *Leptospira* infection reported during the years of this investigation was very close to the results obtained in Northern Italy (11.02%) [38] and in the Liguria region (12.13%) [7]. Moreover, prevalences of 10.30% and 15.3% were found in two different investigations performed in Japan [53,54], while 3.40% was reported in the USA [55]. Based on phylogenetic analysis, pathogenic *Leptospira* DNA in wild boar kidney belong to *L. interrogans*, *L. borgpetersenii*, and *L. kirschneri*. With respect to the serovars in Italy that are more often detected through isolation or serology, [27,38,56–59] and the other serovars that are rarely seropositive [28,60], it might be hypothesized that *L. kirschneri* species found in wild boar kidneys could be related to serogroup Grippotyphosa, while *L. borgpetersenii* species could be related to the serogroups Tarassovi or Ballum. On the other hand, it is very difficult to infer the serogroup related to *L. interrogans* positivity, due to the inclusion of Icterohaemorrhagiae, Canicola, Pomona, Australis, and Sejroe serogroups in this species. Probably, in relation to serological and bacteriological results obtained in this study, the identified *L. interrogans* could belong to serogroup Australis.

The data reported in this investigation suggest that the liver does not seem be a *Leptospira* target organ in wild boar. Furthermore, it could exclude an early stage of infection (leptospiremia) and confirm that positive animals are only chronic renal carriers, as also suggested by isolation from kidneys.

If very little information is available on pathogenic *Leptospira* DNA in wild boar, there is even less data on intermediate *Leptospira*. To the best of these authors' knowledge, it was only in the Liguria region of Italy that 0.49% of wild boar kidneys were positive for intermediate *Leptospira* DNA in the same year of this investigation [7]. As Liguria and Tuscany are two adjoining Italian regions, a large wild boar movement could be a feature of these regions [61–63]. Even though the species of intermediate *Leptospira* from Liguria were not identified, those found in this investigation belong to *L. fainei* species. *L. fainei* was isolated for the first time from fig and was detected in human sera in Australia [64,65] and a human infection with febrile status was reported in France (from a Portuguese citizen) and in two patients in Denmark [66,67]. Considering wild boar behavior and its ability to live in anthropomorphic environment, a transmission between human and wildlife could be possible.

As these are the first determination in European wildlife, more studies are needed to understand the epidemiology of this intermediate *Leptospira* that could causes severe infection in humans [65–67].

The statistical difference presented during the hunting seasons between pathogenic and intermediate *Leptospira* incidence in wild boar could be related to the temperature and the amount of rainfall recorded in Tuscany during these periods. As reported in literature, rainfall and temperature influence the incidence of leptospirosis in humans and animals [12,68–75]. Indeed, from 2018 to 2019, the temperatures and the rainfall were both higher than those from 2019 to 2020 [76–81], suggesting that these atmospheric phenomena could be involved in these seasonality incidence differences.

4. Materials and Methods

4.1. Sample Collection

During two hunting seasons (the first from November 2018 to January 2019 and the second from November 2019 to January 2020) hunted wild boar blood, kidney, and liver were sampled. Blood samples were collected by ocular puncture [82]. The boar's age class was determined after assessing the degree of tooth eruption and the wear and tear of teeth of the lower jaw, considering three age classes: young (under 12 months old), sub-adult (between 12 and 24 months), and adult (over 24 months old). The animal' sex was also recorded [83].

All animals were hunted in the Tuscany region during authorized hunting seasons (November–January), following the regional hunting law (Regolamento di attuazione della legge regionale 12 gennaio 1994, n. 3 D.P.G.R. 48/R/2017). No animals were specifically sacrificed for this study purpose. Animals did not present gross lesions related to infectious disease at postmortem examination, performed during sampling operations.

4.2. Microscopic Agglutination Test (MAT)

Blood samples were centrifugated at 10,000 rpm for 10 minutes to obtain the serum. In order to detect *Leptospira* antibodies, sera were tested through microscopic agglutination test (MAT) [84]. Titer of 1:100 was considered as positive. For the MAT, live *Leptospira* antigens used were: *Leptospira interrogans* serovar Icterohaemorrhagiae (serogroup Icterohaemorrhagiae, strain RGA), *L. interrogans* serovar Canicola (serogroup Canicola, strain Alarik), *L. interrogans* serovar Pomona (serogroup Pomona, strain Mezzano), *L. kirschneri* serovar Grippotyphosa (serogroup Grippotyphosa, strain Moskva V), *L. borgpetersenii* serovar Tarassovi (serogroup Tarassovi, strain Mitis Johnson), *L. interrogans* serovar Bratislava (serogroup Australis, strain Riccio 2), *L. interrogans* serovar Hardjo (serogroup Sejroe, serovar Hardjoprajitno), and *L. borgpetersenii* serovar Ballum (serogroup Ballum, strain Mus 127).

4.3. Leptospira spp. Isolation

Each wild boar organ was cultured in Ellinghausen-McCullough-Johnson-Harris (EMJH) medium (Difco, Detroit, MI, USA). Approximately 10 cm^3 from each organ was homogenized with 5 mL of sterile water and 1 mL of homogenate was cultured in 5 mL of EMJH. Cultures were incubated at 30 °C ± 1 °C for 120 days and observed every 10 days under dark-field microscopy to evaluate possible bacterial growth.

4.4. Molecular Analysis

From each kidney and liver, DNA was extracted using Quick-DNA Plus Kits (Zymo Research, Irvine, CA, USA) following the manufacturer's instructions.

Two different multiplex Realtime-PCR were employed; the first, targeting *Leptospira* spp. (*16S rRNA* gene) and pathogenic *Leptospira* (*lipL32* gene), was performed on all samples [85,86]. The second protocol was only performed on positive *Leptospira* spp. and negative *lipL32* samples, targeting intermediate *Leptospira* (*16S rRNA* gene) and saprophytic *Leptospira* (*23S rRNA* gene) [16,86]. As a positive control for the *lipL32* gene, DNA extracted from a pure culture of *Leptospira interrogans* serogroup Pomona strain Mezzano was used. As a positive control for the 23S rRNA gene for saprophytic *Leptospira*, DNA extracted from a pure culture of *Leptospira biflexa* serogroup Patoc strain Patoc I was used. As a negative control, sterilized saline water was used. A total reaction volume of 15 μL was prepared by using 2x QuantiTect Probe PCR Master Mix (Qiagen, Hilden, Germany), 2 μM of each primer, 500 nM of each probe, and 3 μL of DNA, as previously reported [7]. The RealTime-PCR assay was performed on a Rotorgene Corbett 6000 (Corbett Research, Sydney, Australia) with the following thermal conditions: a holding stage of 95 °C for 5 min and 45 cycles of 95 °C for 15 s and 60 °C for 30 s. Samples with Ct *lipL32* < 35 were considered positive and those samples with 35 < Ct lipL32 ≥ 40 were repeated.

4.5. *Leptospira spp. Characterization and Genotyping*

First, serogroups of the isolates were determined through the MAT using a panel of eight polyclonal anti-sera against the eight serovars reported in Section 4.2. The agglutination with specific antiserum was used to identify the presumptive strain's serogroup [84].

Isolated *Leptospira* were genotyped using a multilocus sequence typing (MLST) scheme based on housekeeping genes [87–89].

Moreover, the *Leptospira* species were identified from positive pathogenic and intermediate *Leptospira* PCR reactions, using primer for *rrs2* gene and *16S rRNA* gene, respectively [86,88].

The amplification of each target gene was realized with HotStarTaq Master Mix Kit (Qiagen, Hilden, Germany), and further sequenced (BMR Genomics, Padova, Italy) using the same amplification primer sets and analyzed using BioEdit Software [90]. Phylogenetic analysis was performed by the maximum likelihood method based on the Tamura–Nei model using MEGA 10 software [91].

4.6. *Statistical Analysis*

Data were analyzed with chi-square (X^2) test. The statistical test was used to evaluate the *Leptospira* infection ratio in relationship to sex (male or female), age class (young, sub-adult, or adult), province (Pisa, Lucca, Livorno, Grosseto, or Siena) and hunting season (2018/2019 or 2019/2020). Statistical significance threshold was set at a p value ≤ 0.05 [92].

5. Conclusions

In conclusion, this investigation confirms through the MAT, isolation, and molecular assays, the role of wild boar in the epidemiology of leptospirosis in Central Italy. Wild boar represents a good indicator of *Leptospira* circulating in a specific area where many different animal species share the same environment. Furthermore, wild boar populations are able to live in a wide spectrum of habitat types, and, have recently reached sub-urban and urban areas. In Italy, little recent data on human leptospirosis are available; however, some studies investigated the prevalence of infection in risk categories (hunters, farmers, and forestry workers) showing serological positivity to *Leptospira* [93,94]. Moreover, on the basis of the most recent report on human leptospirosis in Italy [95], a high infection rate was recorded in adult males, and this could indicate that leptospirosis is strictly related to worker activity. Hunters, for example, are usually all male and over 30 years old. In particular, these peoples are exposed to an high risk of infection due to management and slaughtering of dead animals being performed with little health care [96].

Tarassovi and Bratislava are the two main serogroups that circulate within wild boar in Tuscany. Although Bratislava has been more detected, the isolation of Tarassovi suggests that wild boar could be the main reservoir. In addition, as for pathogenic *Leptospira*, the presence of intermediate species in wild boar kidney underlines the need to perform other studies aimed at understanding the newly-emerging species, *L. fainei*, in animals and in humans.

Author Contributions: Conceptualization, G.C., F.B., and F.F.; investigation, G.C., F.B., and M.A.; data curation, G.C., F.B., M.A., D.C., and F.F.; writing—original draft preparation, G.C., F.B., and F.F.; writing—review and editing, G.C., F.B., M.A., D.C., and F.F.; supervision, D.C. and F.F.; funding acquisition, F.F. All authors have read and agreed to the published version of the manuscript.

Funding: This research was funded by University of Pisa, grant number PRA_2018_56.

Acknowledgments: We thank Mario D'Incau and his team of National Reference Centre for Animal Leptospirosis (Istituto Zooprofilattico Sperimentale della Lombardia e dell' Emilia Romagna "Bruno Ubertini", via Bianchi 7/9, 25121 Brescia, Italy) for provision of the *Leptospira* live strains employed in this investigation.

Conflicts of Interest: The authors declare no conflict of interest.

References

1. Massei, G.; Kindberg, J.; Licoppe, A.; Gačić, D.; Šprem, N.; Kamler, J.; Baubet, E.; Hohmann, U.; Monaco, A.; Ozoliņš, J.; et al. Wild boar populations up, numbers of hunters down? A review of trends and implications for Europe. *Pest Manag. Sci.* **2015**, *71*, 492–500. [CrossRef]
2. Castillo-Contreras, R.; Carvalho, J.; Serrano, E.; Mentaberre, G.; Fernández-Aguilar, X.; Colom, A.; González-Crespo, C.; Lavín, S.; López-Olvera, J.R. Urban wild boars prefer fragmented areas with food resources near natural corridors. *Sci. Total Environ.* **2018**, *615*, 282–288. [CrossRef]
3. Santilli, F.; Varuzza, P. Factors affecting wild boar (Sus scrofa) abundance in southern Tuscany. *Hystrix, Ital. J. Mammal.* **2013**, *24*, 169–173.
4. Pittiglio, C.; Khomenko, S.; Beltran-Alcrudo, D. Wild boar mapping using population-density statistics: From polygons to high resolution raster maps. *PLoS ONE* **2018**, *13*, e0193295. [CrossRef]
5. Carnevali, L.; Pedrotti, L.; Riga, F.; Toso, S. Banca Dati Ungulati: Status, distribuzione, consistenza, gestione e prelievo venatorio delle popolazioni di Ungulati in Italia. Rapporto 2001–2005. *Biol. e Conserv. della Fauna* **2009**, *117*, 1–168.
6. Lombardini, M.; Meriggi, A.; Fozzi, A. Factors influencing wild boar damage to agricultural crops in Sardinia (Italy). *Curr. Zool.* **2017**, *63*, 507–514. [CrossRef] [PubMed]
7. Cilia, G.; Bertelloni, F.; Mignone, W.; Spina, S.; Berio, E.; Razzuoli, E.; Vencia, W.; Franco, V.; Cecchi, F.; Bogi, S.; et al. Molecular detection of Leptospira spp. in wild boar (Sus scrofa) hunted in Liguria region (Italy). *Comp. Immunol. Microbiol. Infect. Dis.* **2020**, *68*, 101410. [CrossRef] [PubMed]
8. Meng, X.J.; Lindsay, D.S.; Sriranganathan, N. Wild boars as sources for infectious diseases in livestock and humans. *Philos. Trans. R. Soc. Lond. B. Biol. Sci.* **2009**, *364*, 2697–2707. [CrossRef] [PubMed]
9. Bertelloni, F.; Mazzei, M.; Cilia, G.; Forzan, M.; Felicioli, A.; Sagona, S.; Bandecchi, P.; Turchi, B.; Cerri, D.; Fratini, F. Serological Survey on Bacterial and Viral Pathogens in Wild Boars Hunted in Tuscany. *Ecohealth* **2020**, *17*, 85–93. [CrossRef] [PubMed]
10. Picardeau, M. Virulence of the zoonotic agent of leptospirosis: Still terra incognita? *Nat. Rev. Microbiol.* **2017**, *15*, 297–307. [CrossRef]
11. Adler, B.; de la Peña Moctezuma, A. Leptospira and leptospirosis. *Vet. Microbiol.* **2010**, *140*, 287–296. [CrossRef] [PubMed]
12. Levett, P.N. Leptospirosis. *Clin. Microbiol. Rev.* **2001**, *14*, 296–326. [CrossRef] [PubMed]
13. Vincent, A.T.; Schiettekatte, O.; Goarant, C.; Neela, V.K.; Bernet, E.; Thibeaux, R.; Ismail, N.; Mohd Khalid, M.K.N.; Amran, F.; Masuzawa, T.; et al. Revisiting the taxonomy and evolution of pathogenicity of the genus Leptospira through the prism of genomics. *PLoS Negl. Trop. Dis.* **2019**, *13*, e0007270. [CrossRef] [PubMed]
14. Guglielmini, J.; Bourhy, P.; Schiettekatte, O.; Zinini, F.; Brisse, S.; Picardeau, M. Genus-wide Leptospira core genome multilocus sequence typing for strain taxonomy and global surveillance. *PLoS Negl. Trop. Dis.* **2019**, *13*, e0007374. [CrossRef]
15. Balamurugan, V.; Gangadhar, N.L.; Mohandoss, N.; Thirumalesh, S.R.A.; Dhar, M.; Shome, R.; Krishnamoorthy, P.; Prabhudas, K.; Rahman, H. Characterization of leptospira isolates from animals and humans: Phylogenetic analysis identifies the prevalence of intermediate species in India. *Springerplus* **2013**, *2*, 362. [CrossRef]
16. Barragan, V.; Chiriboga, J.; Miller, E.; Olivas, S.; Birdsell, D.; Hepp, C.; Hornstra, H.; Schupp, J.M.; Morales, M.; Gonzalez, M.; et al. High Leptospira Diversity in Animals and Humans Complicates the Search for Common Reservoirs of Human Disease in Rural Ecuador. *PLoS Negl. Trop. Dis.* **2016**, *10*, e0004990. [CrossRef]

17. Chiriboga, J.; Barragan, V.; Arroyo, G.; Sosa, A.; Birdsell, D.N.; España, K.; Mora, A.; Espín, E.; Mejía, M.E.; Morales, M.; et al. High Prevalence of Intermediate *Leptospira* spp. DNA in Febrile Humans from Urban and Rural Ecuador. *Emerg. Infect. Dis.* **2015**, *21*, 2141–2147. [CrossRef]
18. Ruiz-Fons, F. A Review of the Current Status of Relevant Zoonotic Pathogens in Wild Swine (*Sus scrofa*) Populations: Changes Modulating the Risk of Transmission to Humans. *Transbound. Emerg. Dis.* **2017**, *64*, 68–88. [CrossRef]
19. Dupouey, J.; Faucher, B.; Edouard, S.; Richet, H.; Kodjo, A.; Drancourt, M.; Davoust, B. Human leptospirosis: An emerging risk in Europe? *Comp. Immunol. Microbiol. Infect. Dis.* **2014**, *37*, 77–83. [CrossRef]
20. Haake, D.A.; Levett, P.N. Leptospirosis in humans. *Curr. Top. Microbiol. Immunol.* **2015**, *387*, 65–97.
21. Ellis, W.A. Animal Leptospirosis. In *Leptospira and Leptospirosis. Current Topics in Microbiology and Immunology*; Adler, B., Ed.; Springer: Berlin, Germany, 2015; Volume 387, pp. 99–137.
22. Cerri, D.; Ebani, V.V.; Fratini, F.; Pinzauti, P.; Andreani, E. Epidemiology of leptospirosis: Observations on serological data obtained by a "diagnostic laboratory for leptospirosis" from 1995 to 2001. *New Microbiol.* **2003**, *26*, 383–389. [PubMed]
23. Żmudzki, J.; Jabłoński, A.; Nowak, A.; Zębek, S.; Arent, Z.; Bocian, Ł.; Pejsak, Z. First overall report of Leptospira infections in wild boars in Poland. *Acta Vet. Scand.* **2015**, *58*, 3. [CrossRef]
24. Vale-Goncalves, H.M.; Cabral, J.A.; Faria, M.C.; Nunes-Pereira, M.; Faria, A.S.; Veloso, O.; Vieira, M.L.; Paiva-Cardoso, M.N. Prevalence of *Leptospira* antibodies in wild boars (*Sus scrofa*) from Northern Portugal: Risk factor analysis. *Epidemiol. Infect.* **2015**, *143*, 2126–2130. [CrossRef] [PubMed]
25. Boqvist, S.; Bergström, K.; Magnusson, U. Prevalence of Antibody to Six Leptospira Servovars in Swedish Wild Boars. *J. Wildl. Dis.* **2012**, *48*, 492–496. [CrossRef] [PubMed]
26. Bertelloni, F.; Cilia, G.; Turchi, B.; Pinzauti, P.; Cerri, D.; Fratini, F. Epidemiology of leptospirosis in North-Central Italy: Fifteen years of serological data (2002–2016). *Comp. Immunol. Microbiol. Infect. Dis.* **2019**, *65*, 14–22. [CrossRef] [PubMed]
27. Tagliabue, S.; Figarolli, B.M.; D'Incau, M.; Foschi, G.; Gennero, M.S.; Giordani, R.; Giordani, R.; Natale, A.; Papa, P.; Ponti, N.; et al. Serological surveillance of Leptospirosis in Italy: Two-year national data (2010-2011). *Vet. Ital.* **2016**, *52*, 129–138.
28. Coppola, F.; Cilia, G.; Bertelloni, F.; Casini, L.; D'Addio, E.; Fratini, F.; Cerri, D.; Felicioli, A. Crested porcupine (Hystrix cristata L.): A new potential host for pathogenic Leptospira among semi-fossorial mammals. *Comp. Immunol. Microbiol. Infect. Dis.* **2020**, *70*, 101472. [CrossRef]
29. Fratini, F.; Turchi, B.; Ebani, V.V.; Bertelloni, F.; Galiero, A.; Cerri, D. The presence of Leptospira in coypus (Myocastor coypus) and rats (Rattus norvegicus) living in a protected wetland in Tuscany (Italy). *Vet. Arh.* **2015**, *85*, 407–414.
30. Bertelloni, F.; Turchi, B.; Vattiata, E.; Viola, P.; Pardini, S.; Cerri, D.; Fratini, F. Serological survey on Leptospira infection in slaughtered swine in North-Central Italy. *Epidemiol. Infect.* **2018**, 1–6. [CrossRef]
31. Ebani, V.V.; Cerri, D.; Poli, A.; Andreani, E. Prevalence of Leptospira and Brucella antibodies in wild boars (Sus scrofa) in Tuscany, Italy. *J. Wildl. Dis.* **2003**, *39*, 718–722. [CrossRef]
32. Verin, R.; Varuzza, P.; Mazzei, M.; Poli, A. Serologic, molecular, and pathologic survey of pseudorabies virus infection in hunted wild boars (Sus scrofa) in Italy. *J. Wildl. Dis.* **2014**, *50*, 559–565. [CrossRef] [PubMed]
33. Ebani, V.V.; Bertelloni, F.; Pinzauti, P.; Cerri, D. Seroprevalence of Leptospira spp. and Borrelia burgdorferi sensu lato in Italian horses. *Ann. Agric. Environ. Med.* **2012**, *19*, 237–240. [PubMed]
34. Hartskeerl, R.; Collares-Pereira, M.; Ellis, W.A. Emergence, control and re-emerging leptospirosis: Dynamics of infection in the changing world. *Clin. Microbiol. Infect.* **2011**, *17*, 494–501. [CrossRef] [PubMed]
35. Montagnaro, S.; Sasso, S.; De Martino, L.; Longo, M.; Iovane, V.; Ghiurmino, G.; Pisanelli, G.; Nava, D.; Baldi, L.; Pagnini, U. Prevalence of Antibodies to Selected Viral and Bacterial Pathogens in Wild Boar (Sus scrofa) in Campania Region, Italy. *J. Wildl. Dis.* **2010**, *46*, 316–319. [CrossRef]
36. Piredda, I.; Palmas, B.; Noworol, M.; Canu, M.; Fiori, E.; Picardeau, M.; Tola, A.; Pintore, A.; Ponti, N. Indagine sulla prevalenza della leptospirosi nei cinghiali del centro-nord Sardegna. In Proceedings of the XIII Congresso Nazionale S.I.Di.L.V., Trani, Italy, 12–14 October 2011; pp. 63–64.
37. Figarolli, B.M.; Gaffuri, A.; Alborali, G.L.; D'Incau, M.; Tagliabue, S. Survey on leptospirosis in wild boars (Sus scrofa) in Lombardy, Northern Italy. In Proceedings of the 1°ELS meeting, Dubrovnik, Croatian, 30 April –1 May 2012; pp. 63–64.

38. Chiari, M.; Figarolli, B.M.; Tagliabue, S.; Alborali, G.L.; Bertoletti, M.; Papetti, A.; D'Incau, M.; Zanoni, M.; Boniotti, M.B. Seroprevalence and risk factors of leptospirosis in wild boars (Sus scrofa) in northern Italy. *Hystrix, Ital. J. Mammal.* **2016**, *27*. [CrossRef]
39. Tagliabue, S.; Raffo, A.; Foni, E.; Candotti, R.; Barigazzi, G. Anticorpi per Leptospira interrogans in sieri di cinghiale selvatico (Sus scrofa) nell'Appennino parmense. In Proceedings of the Convegno Nazionale di Ecopatologia della fauna selvatica, Bologna, Italy, 9–11 February 1995.
40. Espí, A.; Prieto, J.M.; Alzaga, V. Leptospiral antibodies in Iberian red deer (Cervus elaphus hispanicus), fallow deer (Dama dama) and European wild boar (Sus scrofa) in Asturias, Northern Spain. *Vet. J.* **2010**, *183*, 226–227. [CrossRef]
41. Jansen, A.; Nöckler, K.; Schönberg, A.; Luge, E.; Ehlert, D.; Schneider, T. Wild boars as possible source of hemorrhagic leptospirosis in Berlin, Germany. *Eur. J. Clin. Microbiol. Infect. Dis.* **2006**, *25*, 544–546. [CrossRef]
42. Vicente, J.; León-Vizcaíno, L.; Gortázar, C.; Cubero, M.J.; González, M.; Martín-Atance, P. Antibodies to selected viral and bacterial pathogens in European wild boars from southcentral Spain. *J. Wildl. Dis.* **2002**, *38*, 649–652. [CrossRef]
43. Vengust, G.; Lindtner-Knific, R.; Zele, D.; Bidovec, A. Leptospira antibodies in wild boars (Sus scrofa) in Slovenia. *Eur. J. Wildl. Res.* **2008**, *54*, 749–752. [CrossRef]
44. Treml, F.; Pikula, J.; Holešovská, Z. Prevalence of antibodies against leptospires in the wild boar (Sus scrofa L., 1758). *Vet. Med. (Praha).* **2012**, *48*, 66–70. [CrossRef]
45. Arent, Z.; Gilmore, C.; Brem, S.; Ellis, W.A. Molecular studies on European equine isolates of Leptospira interrogans serovars Bratislava and Muenchen. *Infect. Genet. Evol.* **2015**, *34*, 26–31. [CrossRef]
46. Rocha, T.; Ellis, W.A.; Montgomery, J.; Gilmore, C.; Regalla, J.; Brem, S. Microbiological and serological study of leptospirosis in horses at slaughter: First isolations. *Res. Vet. Sci.* **2004**, *76*, 199–202. [CrossRef] [PubMed]
47. Ellis, W.A.; McParland, P.J.; Bryson, D.G.; Cassells, J.A. Boars as carriers of leptospires of the Australis serogroup on farms with an abortion problem. *Vet. Rec.* **1986**, *118*, 563. [CrossRef] [PubMed]
48. Slavica, A.; Cvetnić, Z.; Konjević, D.; Janicki, Z.; Severin, K.; Dežđek, D.; Starešina, V.; Sindičić, M.; Antić, J. Detection of Leptospira spp. serovars in wild boars (Sus scrofa) from continental Croatia. *Vet. Arh.* **2010**, *80*, 247–257.
49. Agampodi, S.B.; Matthias, M.A.; Moreno, A.C.; Vinetz, J.M. Utility of quantitative polymerase chain reaction in leptospirosis diagnosis: Association of level of leptospiremia and clinical manifestations in Sri Lanka. *Clin. Infect. Dis.* **2012**, *54*, 1249–1255. [CrossRef] [PubMed]
50. Merien, F.; Baranton, G.; Perolat, P. Comparison of Polymerase Chain Reaction with Microagglutination Test and Culture for Diagnosis of Leptospirosis. *J. Infect. Dis.* **1995**, *172*, 281–285. [CrossRef]
51. Hall, C.; Lambourne, J. The Challenges of Diagnosing Leptospirosis. *J. Travel Med.* **2014**, *21*, 139–140. [CrossRef] [PubMed]
52. Fornazari, F.; Langoni, H.; Marson, P.M.; Nóbrega, D.B.; Teixeira, C.R. Leptospira reservoirs among wildlife in Brazil: Beyond rodents. *Acta Trop.* **2018**, *178*, 205–212. [PubMed]
53. Koizumi, N.; Muto, M.; Yamada, A.; Watanabe, H. Prevalence of Leptospira spp. in the Kidneys of Wild Boars and Deer in Japan. *J. Vet. Med. Sci.* **2009**, *71*, 797–799. [CrossRef]
54. Koizumi, N.; Uchida, M.; Makino, T.; Taguri, T.; Kuroki, T.; Muto, M.; Kato, Y.; Watanabe, H. Isolation and characterization of Leptospira spp. from raccoons in Japan. *J. Vet. Med. Sci.* **2009**, *71*, 425–429. [CrossRef]
55. Pedersen, K.; Anderson, T.D.; Bevins, S.N.; Pabilonia, K.L.; Whitley, P.N.; Virchow, D.R.; Gidlewski, T. Evidence of leptospirosis in the kidneys and serum of feral swine (*Sus scrofa*) in the United States. *Epidemiol. Infect.* **2017**, *145*, 87–94. [CrossRef] [PubMed]
56. Scanziani, E.; Origgi, F.; Giusti, A.M.; Iacchia, G.; Vasino, A.; Pirovano, G.; Scarpa, P.; Tagliabue, S. Serological survey of leptospiral infection in kennelled dogs in Italy. *J. Small Anim. Pract.* **2002**, *43*, 154–157. [CrossRef] [PubMed]
57. Andreoli, E.; Radaelli, E.; Bertoletti, I.; Bianchi, A.; Scanziani, E.; Tagliabue, S.; Mattiello, S. Leptospira spp. infection in wild ruminants: A survey in Central Italian Alps. *Vet. Ital.* **2014**, *50*, 285–291. [PubMed]
58. Zanzani, S.A.; Cerbo, A.D.; Gazzonis, A.L.; Epis, S.; Invernizzi, A.; Tagliabue, S.; Manfredi, M.T. Parasitic and Bacterial Infections of *Myocastor coypus* in a Metropolitan Area of Northwestern Italy. *J. Wildl. Dis.* **2016**, *52*, 126–130. [CrossRef]

59. Vitale, M.; Agnello, S.; Chetta, M.; Amato, B.; Vitale, G.; Bella, C.D.; Vicari, D.; Presti, V.D.M.L. Human leptospirosis cases in Palermo Italy. The role of rodents and climate. *J. Infect. Public Health* **2018**, *11*, 209–214. [CrossRef] [PubMed]
60. Ciceroni, L.; Lombardo, D.; Pinto, A.; Ciarrocchi, S.; Simeoni, J. Prevalence of antibodies to Leptospira serovars in sheep and goats in Alto Adige-South Tyrol. *J. Vet. Med. B. Infect. Dis. Vet. Public Health* **2000**, *47*, 217–223. [CrossRef]
61. Russo, L.; Massei, G.; Genov, P.V. Daily home range and activity of wild boar in a mediterranean area free from hunting. *Ethol. Ecol. Evol.* **1997**, *9*, 287–294. [CrossRef]
62. Boitani, L.; Mattei, L.; Nonis, D.; Corsi, F. Spatial and Activity Patterns of Wild Boars in Tuscany, Italy. *J. Mammal.* **1994**, *75*, 600–612. [CrossRef]
63. Massolo, A.; Della Stella, R.M. Population structure variations of wild boar *Sus scrofa* in central Italy. *Ital. J. Zool.* **2006**, *73*, 137–144. [CrossRef]
64. Perolat, P.; Chappel, R.J.; Adler, B.; Baranton, G.; Bulach, D.M.; Billinghurst, M.L.; Letocart, M.; Merien, F.; Serrano, M.S. Leptospira fainei sp. nov., isolated from pigs in Australia. *Int. J. Syst. Bacteriol.* **1998**, *48*, 851–858. [CrossRef]
65. Chappel, R.J.; Khalik, D.A.; Adler, B.; Bulach, D.M.; Faine, S.; Perolat, P.; Vallance, V. Serological titres to Leptospira fainei serovar hurstbridge in human sera in Australia. *Epidemiol. Infect.* **1998**, *121*, 473–475. [CrossRef] [PubMed]
66. Arzouni, J.P.; Parola, P.; Scola, B.L.; Postic, D.; Brouqui, P.; Raoult, D. Human infection caused by Leptospira fainei. *Emerg. Infect. Dis.* **2002**, *8*, 865–868. [CrossRef] [PubMed]
67. Petersen, A.M.; Boye, K.; Blom, J.; Schlichting, P.; Krogfelt, K.A. First isolation of Leptospira fainei serovar Hurstbridge from two human patients with Weil's syndrome. *J. Med. Microbiol.* **2001**, *50*, 96–100. [CrossRef] [PubMed]
68. Soares, T.S.M.; do Rosário Dias de Oliveira Latorre, M.; Laporta, G.Z.; Buzzar, M.R. Spatial and seasonal analysis on leptospirosis in the municipality of São Paulo, Southeastern Brazil, 1998 to 2006. *Rev. Saude Publica* **2010**, *44*, 283–291. [CrossRef]
69. Miller, D.A.; Wilson, M.A.; Beran, G.W. Relationships between prevalence of Leptospira interrogans in cattle, and regional, climatic, and seasonal factors. *Am. J. Vet. Res.* **1991**, *52*, 1766–1768.
70. Chadsuthi, S.; Modchang, C.; Lenbury, Y.; Iamsirithaworn, S.; Triampo, W. Modeling seasonal leptospirosis transmission and its association with rainfall and temperature in Thailand using time-series and ARIMAX analyses. *Asian Pac. J. Trop. Med.* **2012**, *5*, 539–546. [CrossRef]
71. Ward, M.P. Seasonality of canine leptospirosis in the United States and Canada and its association with rainfall. *Prev. Vet. Med.* **2002**, *56*, 203–213. [CrossRef]
72. Stoddard, R.A.; Bui, D.; Haberling, D.L.; Wuthiekanun, V.; Thaipadungpanit, J.; Hoffmaster, A.R. Viability of Leptospira isolates from a human outbreak in Thailand in various water types, pH, and temperature conditions. *Am. J. Trop. Med. Hyg.* **2014**, *91*, 1020–1022. [CrossRef]
73. Filho, J.G.; Nazário, N.O.; Freitas, P.F.; Pinto, G.A.; Schlindwein, A.D. Temporal analysis of the relationship between leptospirosis, rainfall levels and seasonality, Santa Catarina, Brazil, 2005-2015. *Rev. Inst. Med. Trop. Sao Paulo* **2018**, *60*, e39.
74. Kupek, E.; de Sousa Santos Faversani, M.C.; de Souza Philippi, J.M. The relationship between rainfall and human leptospirosis in Florianópolis, Brazil, 1991–1996. *Braz. J. Infect. Dis.* **2000**, *4*, 131–134.
75. Hacker, K.P.; Sacramento, G.A.; Cruz, J.S.; De Oliveira, D.; Nery, N.; Lindow, J.C.; Carvalho, M.; Hagan, J.; Diggle, P.J.; Begon, M.; et al. Influence of rainfall on leptospira infection and disease in a tropical urban setting, Brazil. *Emerg. Infect. Dis.* **2020**, *26*, 311–314. [CrossRef] [PubMed]
76. Regione Toscana - Settore Idrologico Regionale. Analisi dei Dati Termometrici Report Anno 2018. 2018. Available online: https://www.sir.toscana.it/supports/download/report/2018_situazione_termometrica.pdf (accessed on 10 May 2020).
77. Regione Toscana - Settore Idrologico Regionale. Report Pluviometrico Anno 2018. 2018. Available online: https://www.sir.toscana.it/supports/download/report/2018_situazione_idrologica.pdf (accessed on 10 May 2020).
78. Regione Toscana - Settore Idrologico Regionale. Analisi dei Dati Termometrici Report Anno 2019. 2019. Available online: https://www.sir.toscana.it/supports/download/report/2019_situazione_termometrica.pdf (accessed on 10 May 2020).

79. Regione Toscana - Settore Idrologico Regionale. Report Pluviometrico Anno 2019. 2019. Available online: https://www.sir.toscana.it/supports/download/report/2019_situazione_idrologica.pdf (accessed on 10 May 2020).
80. Regione Toscana - Settore Idrologico Regionale. Report Pluviometrico del Mese di Gennaio 2020. 2020. Available online: https://www.sir.toscana.it/supports/download/report/2020-01_report_cumulate_mensili.pdf (accessed on 10 May 2020).
81. Regione Toscana - Settore Idrologico Regionale. Analisi Dati Termometrici Report Gennaio 2020. 2020. Available online: https://www.sir.toscana.it/supports/download/report/report_termometria_2020-01.pdf (accessed on 10 May 2020).
82. Arenas-Montes, A.; García-Bocanegra, I.; Paniagua, J.; Franco, J.J.; Miró, F.; Fernández-Morente, M.; Carbonero, A.; Arenas, A. Blood sampling by puncture in the cavernous sinus from hunted wild boar. *Eur. J. Wildl. Res.* **2013**, *59*, 299–303. [CrossRef]
83. Sáez-Royuela, C.; Gomariz, R.P.; Luis Tellería, J. *Age Determination of European Wild Boar*; Wiley: Hoboken, NJ, USA, 1989; Volume 17.
84. OIE. Leptospirosis. *Man. Diagnostic Tests Vaccines Terr. Anim.* **2014**. Available online: https://www.oie.int/fileadmin/Home/eng/Health_standards/tahm/3.01.12_LEPTO.pdf (accessed on 10 May 2020).
85. Stoddard, R.A.; Gee, J.E.; Wilkins, P.P.; McCaustland, K.; Hoffmaster, A.R. Detection of pathogenic Leptospira spp. through TaqMan polymerase chain reaction targeting the LipL32 gene. *Diagn. Microbiol. Infect. Dis.* **2009**, *64*, 247–255. [CrossRef]
86. Bedir, O.; Kilic, A.; Atabek, E.; Kuskucu, A.M.; Turhan, V.; Basustaoglu, A.C. Simultaneous detection and differentiation of pathogenic and nonpathogenic Leptospira spp. by multiplex real-time PCR (TaqMan) assay. *Polish J. Microbiol.* **2010**, *59*, 167–173. [CrossRef]
87. Boonsilp, S.; Thaipadungpanit, J.; Amornchai, P.; Wuthiekanun, V.; Bailey, M.S.; Holden, M.T.G.; Zhang, C.; Jiang, X.; Koizumi, N.; Taylor, K.; et al. A single Multilocus Sequence Typing (MLST) scheme for seven pathogenic Leptospira species. *PLoS Negl. Trop. Dis.* **2013**, *7*, e1954. [CrossRef] [PubMed]
88. Ahmed, N.; Devi, S.M.; de los Á Valverde, M.; Vijayachari, P.; Machang'u, R.S.; Ellis, W.A.; Hartskeerl, R.A. Multilocus sequence typing method for identification and genotypic classification of pathogenic Leptospira species. *Ann. Clin. Microbiol. Antimicrob.* **2006**, *5*, 28. [CrossRef] [PubMed]
89. Varni, V.; Ruybal, P.; Lauthier, J.J.; Tomasini, N.; Brihuega, B.; Koval, A.; Caimi, K. Reassessment of MLST schemes for Leptospira spp. typing worldwide. *Infect. Genet. Evol.* **2014**, *22*, 216–222. [CrossRef] [PubMed]
90. Hall, T.A. BioEdit: A user-friendly biological sequence alignment editor and analysis program for Windows 95/98/NT. *Nucleic Acids Symp. Ser.* **1999**, *41*, 95–98.
91. Kumar, S.; Stecher, G.; Li, M.; Knyaz, C.; Tamura, K. MEGA X: Molecular Evolutionary Genetics Analysis across Computing Platforms. *Mol. Biol. Evol.* **2018**, *35*, 1547–1549. [CrossRef]
92. R. R Development Core Team. A language and environment for statistical computing. *R Found. Stat. Comput. Vienna, Austria* **2015**. [CrossRef]
93. Nuti, M.; Amaddeo, D.; Crovatto, M.; Ghionni, A.; Polato, D.; Lillini, E.; Pitzus, E.; Santini, G.F. Infections in an alpine environment: Antibodies to hantaviruses, leptospira, rickettsiae, and Borrelia burgdorferi in defined Italian populations. *Am. J. Trop. Med. Hyg.* **1993**, *48*, 20–25. [CrossRef]
94. Nuti, M.; Amaddeo, D.; Autorino, G.L.; Crovatto, M.; Crucil, C.; Ghionni, A.; Giommi, M.; Salvati, F.; Santini, G.F. Seroprevalence of antibodies to hantaviruses and leptospires in selected Italian population groups. *Eur. J. Epidemiol.* **1992**, *8*, 98–102. [PubMed]
95. Istisan, R.; Graziani, C.; Duranti, A.; Morelli, A.; Busani, L.; Pezzotti, P. Rapoorti ISTISAN Istituto Superiore di Sanità 16/1. In Proceedings of the I Simposio internazionale Nuove strategie per gli interventi di prevenzione dello stress da lavoro, Sassari-Alghero, Italy, 8–10 July 2015.
96. Ngugi, J.N.; Fèvre, E.M.; Mgode, G.F.; Obonyo, M.; Mhamphi, G.G.; Otieno, C.A.; Cook, E.A.J. Seroprevalence and associated risk factors of leptospirosis in slaughter pigs; A neglected public health risk, western Kenya. *BMC Vet. Res.* **2019**, *15*, 403. [CrossRef] [PubMed]

© 2020 by the authors. Licensee MDPI, Basel, Switzerland. This article is an open access article distributed under the terms and conditions of the Creative Commons Attribution (CC BY) license (http://creativecommons.org/licenses/by/4.0/).

pathogens

MDPI

Article

Laboratory Diagnosis of Bovine Abortions Caused by Non-Maintenance Pathogenic *Leptospira* spp.: Necropsy, Serology and Molecular Study Out of a Belgian Experience

Fabien Grégoire [1], Raïssa Bakinahe [2], Thierry Petitjean [1], Samira Boarbi [2,†], Laurent Delooz [1], David Fretin [2], Marc Saulmont [1] and Marcella Mori [2,*]

1 Association Régionale de Santé et d'Identification Animales (ARSIA), B-5590 Ciney, Belgium; fabien.gregoire@arsia.be (F.G.); thierry.petitjean@arsia.be (T.P.); laurent.delooz@arsia.be (L.D.); marc.saulmont@arsia.be (M.S.)

2 Bacterial Zoonoses of Animals Unit, Veterinary Bacteriology, Sciensano, 1180 Brussels, Belgium; Raissa.BakinaheNtamukunzi@sciensano.be (R.B.); Samira.Boarbi@sciensano.be (S.B.); David.Fretin@sciensano.be (D.F.)

* Correspondence: Marcella.Mori@sciensano.be

† Present affiliation: Belgian Reference Laboratory for Tuberculosis and Mycobacteria, Sciensano, 1180 Brussels, Belgium.

Received: 24 April 2020; Accepted: 22 May 2020; Published: 26 May 2020

check for updates

Abstract: Bovine leptospirosis is a bacterial zoonotic disease caused by pathogenic *Leptospira* spp. The pathology and epidemiology of this infection are influenced by the numerous existing serovars and their adaptation to specific hosts. Infections by host-maintained serovars such as Hardjo are well documented, unlike those from the incidental ones. In July 2014, an emerging phenomenon of an increased incidence of icteric abortions associated with leptospiral infection occurred in southern Belgium. First-line serological analyses targeting cattle-adapted serovars failed at initial diagnosis. This study provides a comprehensive description of laboratory findings—at the level of necropsy, serology and molecular diagnosis—regarding icteric and non-icteric abortions (n = 116) recorded during this time (years 2014–2015) and associated with incidental infection by serovars such as Grippotyphosa, Australis and Icterohaemorrhagiae. Based on these tests, a diagnostic pathway is proposed for these types of infection in cattle to establish an affordable but accurate diagnosis in the future. These investigations add insights into the understanding of the pathogenesis of bovine leptospirosis associated with serovars classically described as non-maintenance.

Keywords: *Leptospira* spp.; cattle; abortion; pathology; non-maintenance serovars; MAT; PCR; *lfb1*-phylogeny

1. Introduction

Bovine leptospirosis is a bacterial zoonotic disease caused by pathogenic *Leptospira* spp., bacteria classified in about 300 serovars and 64 genomospecies, of which 37 belong to the pathogenic clade [1–3]. The pathology and epidemiology of this bacterial disease are influenced by the existing serovars and their adaptation to specific hosts in wildlife or livestock [1]. Depending on the serovar–host relationship, incidental and host-maintained infections are described.

Leptospiral infections in cattle are known as incidental when involving non-maintenance pathogenic *Leptospira* spp. such as the serogroups Icterohaemorrhagiae, Hebdomadis, Grippotyphosa and Canicola [4–6]. These infections are transmitted through indirect or, more rarely, direct contact with an infected host, generally a wildlife reservoir (i.e., rodents for the serogroup Icterohaemorrhagiae [7,8]). On the other hand, cattle can act as the principal reservoir of the bacteria for the adapted *L. borgepetersenii* serovar Hardjobovis and *L. interrogans* serovar Hardjoprajitno. In this latter case, the transmission is more efficient and mainly occurs through direct contact with the contaminated urine or body fluids (milk or placental fluids) of infected animals residing in the same herd [9].

Infection by non-maintenance pathogenic *Leptospira* spp. is characterized by an acute form of the disease such as a high fever, hemolytic anemia, jaundice and multi-systemic illness [10]. Acute manifestation is rarely described for maintenance strains such as *L. borgepetersenii* serovar Hardjobovis and *L. interrogans* serovar Hardjoprajitno [11–14]. In this case, the host presents a sub-acute clinical phase of the disease that often goes unobserved. Bovine-adapted serovars follow a chronic course of infection that is associated with a prolonged renal carrier state of the adult animal associated with chronic renal diseases [15–17]. Chronically infected animals can excrete leptospires through urine, leading to increased environmental infection pressure. Economic issues can be important since these infections also cause reproductive failure, abortions, stillbirths, weak offspring and reduced milk production. Hence, it is a concern for public and animal health management. The management and diagnosis of infections due to *L. borgepetersenii* serovar Hardjobovis and *L. interrogans* serovar Hardjoprajitno in cattle are supported by numerous experimental and epidemiological studies [18–24]. *L. borgepetersenii* serovar Hardjobovis and *L. interrogans* serovar Hardjoprajitno have been shown to be causes of reproductive losses worldwide [25–27]. Instead, knowledge of the disease caused by non-maintenance serovars, other than from serological surveillance, is poor and linked to anecdotal reports [28–30]. The pathogenesis of reproductive disease due to *Leptospira* spp. remains poorly understood as well. Transplacental infection, occurring during the very limited time of maternal bacteremia, is supposed to be the sole cause of abortion [31].

In July 2014, an emerging phenomenon associated with an increased incidence of icteric abortions occurred in the southern region of Belgium [32]. Initial laboratory findings highlighted in dams serum high-level antibody titers against *Leptospira* serogroups Grippotyphosa and Australis, whilst a first-line ELISA-based serologic survey for *L. borgepetersenii* serovar Hardjobovis failed to identify the causative agent [32]. A definitive laboratory diagnosis was finally achieved at the animal national reference leptospirosis laboratory in the fall of 2014 by systematically associating positive serology in dams with the detection of pathogenic *Leptospira* spp. DNA in various organs of the abortuses. The conclusions were promptly reported to the national competent authorities (official national internal report). One epidemiological study thereafter conducted on prospective samples confirmed serological reactivity against *Leptospira* serogroups Grippotyphosa and Australis and provided genotypic indication in one sample of the same year, 2014 [33].

This study provides a comprehensive description of laboratory findings— at the level of necropsy, serology and molecular characterization—regarding the icteric and non-icteric abortions recorded during the years 2014–2015 and associated with leptospiral infection. Based on these tests and their relative intrinsic qualities, the goal of this work is to provide comprehensive tools for the clinical and laboratory diagnosis of bovine leptospirosis due to non-maintenance serovars. Besides, a diagnostic pathway is proposed to establish an affordable but accurate diagnosis of these cases in the future. Finally, these investigations add insights into the understanding of the pathogenesis and clinical manifestation of bovine leptospirosis associated with serovars classically described as non-maintenance.

2. Results

2.1. Characteristics at Necropsy of MAT Positive Abortions

A total of 116 fetuses were selected for the entire study. The gestational age varied from 3 to 9 months, with the majority of abortions occurring during the last trimester of gestation (Table 1). When considering the lesions assessed during necropsy, four entries were initially selected due to previously being observed in icteric abortions [32]: icterus (n = 52), splenomegaly (n = 59), coppery liver (n = 40) and peri-renal hemorrhage (n = 59). However, these entries accounted only partially for the microscopic agglutination test (MAT)-positive results observed (Table 2 and Table S1). Therefore, the presence of generalized hemorrhagic edema (n = 6) was additionally included to better understand the complete lesion panel associated with *Leptospira* infection. For each of these lesions, the association with a positive MAT result (regardless of the reactive serogroup) was evaluated. Four different seropositivity cut-offs for MAT were considered: 1/10, 1/100, 1/300 and 1/1000 (Table 2). A significant association was observed between seropositivity and the presence of icterus, splenomegaly or coppery liver for all of the considered cut-offs. A similar observation was made for the combination of these three lesions. By contrast, no link was established for the peri-renal hemorrhages, regardless of the MAT cut-off. An extended hemorrhagic pattern was associated with a positive MAT result for the cut-off of 1/1000 and, interestingly, observed only in the non-icteric abortion group (Table 2).

Table 1. Distribution of the autopsied fetuses according to their gestational age, presence of icterus (other lesions not included) and positivity rates of MAT (cut-off of 1/10).

Month of Abortion	N Autopsied	Icteric (N Positive/ N Analyzed for Lesions)	Positive MAT in Dam's Sera (N Positive/N Tested)	Positive MAT in Pleural Fluids (N Positive/N Tested)	Positive PCR (N Positive/N Tested)
<5	8	0/8	5/7	0/1	0/0
5	2	0/2	1/2	0/0	0/1
6	5	0/5	4/5	0/0	0/2
7	7	2/6	4/7	0/0	0/5
8	43	18/40	30/42	1/9	13/25
9	51	32/50	31/45	0/16	19/37
TOTAL	**116**	**52/111**	**75/108**	**1/26**	**32/70**

Table 2. Association between necropsy findings in bovine abortuses and results of MAT (positive reaction for any serogroup) performed on sera of the corresponding dams. P = presence; A = absence.

Lesion		N Fetuses	Positivity Cut-Off of MAT							
			1/10		1/100		1/300		1/1000	
			% Positive MAT (n)	OR (95% IC) χ^2 and p-Value	% Positive MAT (n)	OR (95% IC) χ^2 and p-Value	% Positive MAT (n)	OR (95% IC) χ^2 and p-Value	% Positive MAT (n)	OR (95% IC) χ^2 and p-Value
Icterus	P	44	90.91 (40)	12.61 (3.94–40.40) $\chi^2 = 20.99$ $p < 0.001$	88.64 (39)	13.55 (4.56–40.24) $\chi^2 = 24{,}91$ $p < 0.001$	77.27 (34)	14.28 (5.33–38.29) $\chi^2 = 30{,}04$ $p < 0.001$	43.18 (19)	7.14 (4.09-12.48) $\chi^2 = 12.59$ $p < 0.001$
	A	52	44.23 (23)		36.54 (19)		19.23 (10)		9.62 (5)	
Splenomegaly	P	54	83.33 (45)	6.67 (2.60–17.09) $\chi^2 = 15.41$ $p < 0.001$	77.78 (42)	5.69 (2.33–13.91) $\chi^2 = 13{,}94$ $p < 0.001$	61.11 (33)	4.43 (1.84–10.67) $\chi^2 = 10.24$ $p = 0.001$	33.33 (18)	3.00 (1.12–8.04) $\chi^2 = 3.61$ $p = 0.057$
	A	42	42.86 (18)		38.10 (16)		26.19 (11)		14.29 (6)	
Coppery liver	P	36	86.11 (31)	5.43 (1.86–15.85) $\chi^2 = 9.31$ $p = 0.002$	80.56 (29)	4.43 (1.68–11.66) $\chi^2 = 8.47$ $p = 0.004$	69.44 (25)	5.99 (2.35–15.29) $\chi^2 = 13.64$ $p < 0.001$	44.44 (16)	5.20 (2.83–9.56) $\chi^2 = 10.01$ $p = 0.002$
	A	60	53.33 (32)		48.33 (29)		31.67 (19)		13.33 (8)	
Icterus + splenomegaly + coppery liver	P	29	89.66 (26)	7.03 (1.94–25.49) $\chi^2 = 9.16$ $p = 0.002$	86.21 (25)	6.44 (2.02–20.52) $\chi^2 = 10.06$ $p = 0.002$	75.86 (22)	6.43 (2.38–17.33) $\chi^2 = 13.41$ $p < 0.001$	44.83 (13)	4.14 (2.46–6.97) $\chi^2 = 7.26$ $p = 0.007$
	A	67	55.22 (37)		49.25 (33)		32.84 (22)		13.33 (11)	
Peri-renal hemorrhages	P	54	64.81 (35)	0.92 (0.39–2.16) $\chi^2 = 0.0007$ $p = 0.978$	57.41 (31)	0.75 (0.33–1.72) $\chi^2 = 0.22$ $p = 0.636$	44.44 (24)	0.88 (0.39–1.98) $\chi^2 = 0.01$ $p = 0.918$	25.93 (14)	1.12 (0.44–2.85) $\chi^2 = 0.007$ $p = 0.934$
	A	42	66.67 (28)		64.29 (27)		47.62 (20)		23.81 (10)	
Extended hemorrhagic pattern	P	6	83.33 (5)	2.76 (0.31–24.65) $p = 0.661$	83.33 (5)	3.49 (0.39–31.12) $p = 0.397$	83.33 (5)	6.54 (0.73–58.26) $p = 0.09$	66.67 (4)	7.00 (1.19–41.04) $p = 0.033$
	A	90	64.44 (58)		58.89 (53)		43.33 (39)		22.22 (20)	
TOTAL		96	65.63 (63)		60.42 (58)		45.83 (44)		25.00 (24)	

2.2. Serological Pattern of Abortions

Serological analyses were performed on 88 individual dams' sera, on 20 coupled dams' sera and fetuses' pleural fluids, and six individual fetuses' pleural fluids. Pleural fluids were chosen as an alternative to missing post-mortem fetal sera because they reflect the serum antibody levels arising through passive diffusion [34]. These analyses allowed the obtaining of MAT profiles for all but two abortuses. Globally, a higher positivity rate was observed in abortuses with icteric (42/50, 80.7%) vs. non-icteric (33/64, 51.5%) patterns (Table S1). When both dam's serum and fetus' pleural fluid were tested and a positive serological MAT result was observed (18/20 abortions), positivity was observed only in the dam's serum in 17 out of 18 cases (94.4%) (Table S1). For one single fetus, serologically positive reactions were observed in both the dam (high titer) and the pleural fluid of the fetus (low titer), with perfect agreement on the presumptive serogroup. The presumptive infecting *Leptospira* serogroup was defined in 65/75 (86.7%) of the MAT-positive abortuses, with the following frequencies: Grippotyphosa (n = 42), Australis (n = 13), Ballum (n = 4), Icterohaemorrhagiae (n = 2), Autumnalis (n = 1), Bataviae (n = 1), Pyrogenes (n = 1) and Tarassovi (n = 1). None of these serogroups belongs to the pathogenic *Leptospira* serovars classically described as maintenance in cattle. Cross-reactivities between the serogroups were observed, in order of decreasing frequency, between the serogroups Grippotyphosa and Ballum (18/75), Grippotyphosa and Autumnalis (14/75), Grippotyphosa and Australis (10/75), and Australis and Autumnalis (9/75) (Table S2).

2.3. Bacteriological Characteristics of Abortions

PCR was performed on 68 fetuses in available organs. A fetus was considered positive when a positive PCR result was reported in at least one organ. Again, clear high positivity rates were observed in abortuses with icteric (30/41, 73.17%) vs. non-icteric (1/27, 3.70%) patterns, as shown in Table 3. When associations were calculated between each lesion and a positive PCR result in abortuses, a significant link was only found for icterus, splenomegaly, coppery liver and the combination of these three lesions. No significant link was observed for peri-renal hemorrhages and general hemorrhagic pattern (Table 3). Based on the PCR results in the different organs, the placenta, spleen and liver appeared as the most appropriate of a fetus' organs in which to look for non-maintenance *Leptospira* spp. colonization (Table 4). The positive rate was significantly higher in the placenta than in the kidney (χ^2 = 6.15, p-value = 0.013). The median Ct (cycle threshold) values were averagely high in all the organs, and most of the Cts were above 30. For all but five of the tested fetuses (68), the MAT results for the dam's serum were available to define the possible relations between PCR and MAT tests (Table S3). Cohen's kappa coefficient, as well as the relative sensitivity and specificity towards PCR, were calculated for each cut-off of the MAT test. The cut-off of 1/300 showed the highest concordance value (kappa = 0,679, substantial agreement), with a relative sensitivity and a specificity approaching 85% (Table S3).

Table 3. Association between necropsy findings in bovine abortuses and PCR results for the diagnosis of leptospirosis. P = presence; A = absence.

Lesions		N Fetuses	% Positive PCR (n)	OR (95% IC) χ^2 and p-Value
Icterus	P	41	73.17 (30)	70.91 (8.57–586.89) χ^2= 31.67 (p < 0.001)
	A	27	3.70 (1)	
Splenomegaly	P	45	64.44 (29)	19.03 (3.95–91.81) χ^2= 16.89 (p < 0.001)
	A	23	8.70 (2)	
Coppery liver	P	31	70.97 (22)	7.60 (2.58–22.38) χ^2= 12.97 (p < 0.001)
	A	37	24.32 (9)	
Icterus + splenomegaly + coppery liver	P	28	75.00 (21)	9.00 (2.95–27.45) χ^2= 14.65 (p < 0.001)
	A	40	25.00 (10)	
Peri-renal hemorrhages	P	49	46.94 (23)	1.22 (0.42–3.54) χ^2= 0.008 (p =0.930)
	A	19	42.11 (8)	
Extended hemorrhagic pattern	P	3	33.33 (1)	0.58 (0.05-6.76) p = 1
	A	65	46.15 (30)	
TOTAL		**68**	**45.59 (31)**	

Table 4. PCR results obtained in various organs of the autopsied fetuses* (n = 68).

Organ	N Positive/N Tested	Positivity (%)	Median Ct Value	Range Min–Max	
Spleen	18/45	40.00	36.67	31.06	41.45
Placenta [a]	17/32	53.13	34.14	27.52	41.41
Kidney [a]	4/21	19.05	33.92	28.10	38.64
Adrenal glands	1/15	6.67	36.60	36.60	36.60
Liver	5/14	35.71	34.58	31.90	37.60
Lung	2/10	20.00	34.61	32.30	36.92
Brain	0/5	0.00	/	/	/
Hepatic lymph nodes	2/3	66.67	37.45	36.70	38.20

* One abortus might be positive for several organs. [a] significantly different (p = 0.013).

2.4. Molecular Typing of *Leptospira spp.*

To better understand the epidemiology of *Leptospira*-induced abortions in this study and to identify, at the species level, the *Leptospira* infecting strain, the analysis of the sequence polymorphism in the fibronectin-binding protein gene (*lfb1*) gene was performed by using both high-resolution melting analysis (HRMA) and amplicon sequencing. The melting curve analysis of eight fetuses indicated the presence of two main types of PCR amplicon, with different melting temperatures (Tm), highlighting the presence of at least two *Leptospira* species causing abortions. The first cluster, with an average 81.8 °C Tm, was assigned, by comparison with the Tm of reference strains, to the *Leptospira interrogans* group; the second cluster, with a one-degree difference in Tm (average of 82.8 °C) was assigned to the *Leptospira kirschneri* group (Figure S1). Phylogenetic analyses of the *lfb1* sequence in seven fetuses and reference strains (ours or those available in databases) corroborated observations by HRMA. The results, supported by the amplification of a minimum of 238 bp of sequence (accession numbers KY373222 to KY373229), evidenced the heterogeneity of the species involved in abortions: three *Leptospira interrogans* clusters (1a, 1b and 2) including the serovars Australis Ballico, Autumnalis and Hardjoprajitno (cluster 1a); the serovars Australis Bratislava and Bataviae (cluster 1b); the serovars Pyrogenes and Copenhageni (cluster 2); one *Leptospira kirschneri* cluster including the reference sequences of the serovars Cynopteri and Grippotyphosa (Figure 1). Except for the fetus f044, where the presumptive serogroup could not be identified by MAT, the *Leptospira* genospecies based on molecular typing is compatible with the presumptive serogroups determined by MAT in dams (Table 5).

Table 5. Comparison of results in autopsied fetuses according to high-resolution melting analysis (HRMA), *lfb1* phylogeny and serology in dams.

		HRMA rtPCR		*lfb1* Sequencing	Serology of Dam	Necropsy
Fœtus ID	**Sample ID**	**Tm**	**Species Group Attribution**	**Species Group Attribution**	**MAT**	**Group**
f050	15095924-1509927-159915	81.92	*interrogans*	*interrogans* cluster 1b	1/500 Australis	Icteric
f033	14040522.1-1610225.2	81.93	*interrogans*	*interrogans* cluster 1a	1/1000 Icterohaemorrhagiae	Icteric
f010	14048419.1-1610225.3	81.69	*interrogans*	*interrogans* cluster 2	1/500 Australis	Icteric
f032	14045546	81.71	*interrogans*	*interrogans* cluster 1b	-	Icteric
f044	14057925-1417554.1-1417819	81.72	*interrogans*	*interrogans* cluster 1b	ND (1/100 Ballum 1/100 Grippothyphosa)	Icteric
f028	15050216-1506066.2	82.96	*kirschneri*	*kirschneri* cluster	1/100 Grippotyphosa	Icteric
f007	14047394.1-1610225.1	82.82	*kirschneri*	-	1/500 Grippotyphosa	Icteric
f031	14044550	82.76	*kirschneri*	*kirschneri* cluster	-	Icteric

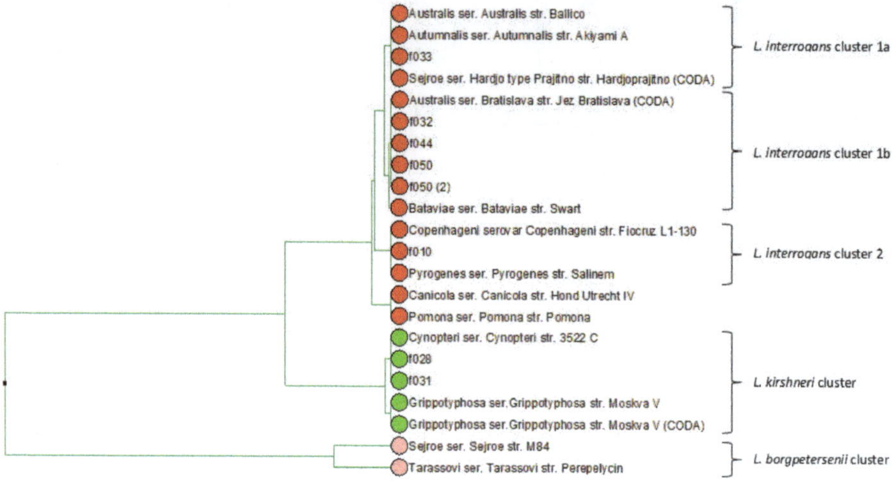

Figure 1. *lbf1*-derived phylogeny of Leptospira-induced icteric abortions and reference strains. Red, green and rose circles indicate strains of *L. interrogans*, *L.kirschneri* and *L. borgepetersenii* species, respectively. Reference strains are listed with their serogroup, serovar and strain name. Abortuses are indicated with their fetus ID as in Table S1.

3. Discussion

Abortion is a well-known manifestation of leptospirosis in cattle. Nevertheless, extended anatomo-pathological descriptions of fetuses infected with leptospires are rarely available, especially for serovars other than *L. borgepetersenii* serovar Hardjobovis. The emergence of abortions due to non-maintenance leptospires observed in 2014 in southern Belgium showed that general icterus was the most striking lesion reported in the abortuses [32]. Jaundice is an acute manifestation commonly reported in newborn calves infected by *Leptospira* spp., due to hemolysis following hemolysin production. This lesion was also reported in fetuses in cases of experimental abortions with *L. borgepetersenii* serovar Hardjobovis [35]. On the basis of the abortuses included in this study, coppery liver and splenomegaly, besides icterus, were significantly associated with a positive result for *Leptospira*, for both serology and bacteriology. These lesions in abortuses are not pathognomonic, and their appreciation depends on the state of conservation of the cadaver. Less distinct lesions such light icterus could be missed by an unaware or untrained eye. Despite these minor drawbacks, necropsy of the fetuses remains, as strongly confirmed in this study, an efficient tool in the global context of abortion surveillance. As of the Belgian experience, such monitoring proved its efficacy during the re-emergence episodes of the Schmallenberg virus and, to a lesser extent, brucellosis [36,37]. In addition to the three mentioned lesions, generalized hemorrhagic edema has been observed in some abortuses and was shown to be associated with high antibody titers in dams. No specific serovar was associated with this lesion. Petechial and ecchymotic hemorrhages on internal organs completed the picture in most of these cases, whereas jaundice was absent in these fetuses except in one. The high *Leptospira* titers associated with generalized hemorrhagic edema suggested a per-acute phenomenon in this case. This feature might correspond to the first stage of the disease leading to an icteric pattern or even represent a completely different pathogenesis of bovine leptospirosis. Indeed, the outcome of *Leptospira* infection varies depending on the infecting serovars but also on host specificities or host–pathogen interaction [38].

Globally, serological analyses indicated that non-maintenance *Leptospira* spp. serogroups were, in different proportions, the causes of the icteric-hemorrhagic abortions observed in this study, namely serogroups Grippotyphosa, Australis, Ballum, Icterohaemorrhagiae, Autumnalis, Bataviae,

Pyrogenes and Tarassovi. The seropositive results could not be attributed to vaccination, as a vaccine for cattle is not marketed in Belgium. The comparison of our results with European trends is difficult, although our data are in line with the marked increase in leptospirosis infection observed in the year 2014 [39,40]. However, a complete epidemiological picture of *Leptospira* spp. serogroups in cattle across Europe is unavailable. The reasons for this are various; seroprevalence studies are often restricted to a specific timeframe, data are not centralized at the European level for animals and the standardization of the MAT analyses is subject to constraints. The Grippotyphosa, Australis, Icterohaemorrhagiae and—to a lesser extent—Autumnalis and Bataviae serogroups have been observed in clinically relevant cattle herds in France [41]. Although this in line with our data, information is missing to ascertain whether the reproductive disorders described in France were similar to the ones observed in our study. Seroprevalence studies performed in cattle herds in Italy show that the non-maintenance serogroups Australis and Icterohaemorrhagiae are observed at high titer levels, together with the maintenance serovars Hardjo and Pomona, while Ballum, Canicola, Grippotyphosa and Tarassovi are rare and at minimal titers [42]. In Spain, the relevant cattle serovars are Bratislava, Hardjo and Pomona [43,44]. Limited seroprevalence data are present in Belgium to understand the relationship between the serogroups found in this study and their possible presence/spread in other (reservoir) animals; Australis and Grippotyphosa are the main serogroups found in dogs [45], and they were also detected in muskrats [34,46]. In humans, a marked increase in autochthonous cases was reported in 2014 [47], as observed in other parts of Europe [39], but a univocal association of the serovars involved in bovine icteric abortions in 2014 with human cases could not be established (National Reference Laboratory personal communication).

Molecular diagnosis, supported by serology, highlighted two mains genospecies responsible for abortions in this study: *L. interrogans* and *L. kirschneri*. Although this was relatively clear from our previous investigations [32], here, we demonstrated by *lfb1* phylogeny that infections could be sourced back to genetically diverse strains (for instance, at least three clusters for the *L. interrogans* species) and not to a unique "outbreak" clone. This could probably explain the heterogenous serological response against *Leptospira* agglutinins and variation observed with the MAT analyses. Although the *L. interrogans* cluster included sequences that were very closely related to the *L. interrogans* serovar Hardjoprajitno, the involvement of this serovar was excluded in our conclusion as it was not supported by serological analyses operated both by ELISA [32] and MAT.

Together with the above observations allowing a better understanding of the *Leptospira* pathogenesis in cattle, this study provided additional insights for the laboratory diagnosis of bovine leptospirosis. First, antibodies against *Leptospira* spp. were detected by MAT in the dam's serum and not in the pleural fluid of the fetuses, despite positive PCR results for fetus organs. This indicated that the dam's serum was the most appropriate sample type for serodiagnosis in this study. Although we previously demonstrated that pleural fluids, in absence of post-mortem sera, are informative for leptospiral infective status [34], these samples taken from the bovine fetus are possibly not relevant in case of cattle infection with leptospires. It can also be speculated that we could not identify seropositive reactions because the fetuses were in bacteremia, which is associated in humans with an absent or very low serological antibody response [48]. Second, the cut-off for MAT was lowered to as low as 1/10 dilution. In classical serological diagnosis, the MAT cut-off is kept at 1/100 (or at 1/400 for higher specificity); the OIE manual suggests lowering this threshold in the case of serosurveillance studies [49]. We suggest lowering this threshold also in case of clustered cases recorded in a defined time-lapse and linked to unusual phenomena. Third, cross-reactivities between serovars were likely to occur, and, with the exception of few cases, they did not hamper the final main serogroup presumptive identification, which was in high agreement with the genospecies attribution. Fourth, leptospiral antigens in abortuses were found in additional organs other than the placenta, kidneys, liver and adrenal glands [50]. As suggested by our study, *Leptospira* spp. are likely to be also detected by PCR in the spleen. This is not surprising since this organ acts like a blood filter and thus is susceptible to sequestering bacteria. Last, this study strongly supports, in the case of poorly loaded DNA samples, the use of *lfb1* phylogeny with the HRMA or sequencing methodology for outbreak-clustered case investigations to provide enough discriminatory genetic power for difficult samples [51,52].

Some limitations could be present in our study. A positive PCR result in the abortus indicates that the bacteria reached the gravid uterus, suggesting a strong link between the presence of *Leptospira* and abortion. However, most of the Ct results from PCR were high, suggesting a low quantity of bacteria in fetal tissues, and some cases could have be missed due to the limit of detection of the method. In addition, in our study, a negative PCR result could not allow the exclusion of *Leptospira* from the abortion causes because of the non-systematic testing of all the organs. For example, the placenta was not systematically transmitted with the fetus for analysis. Additionally, a defect in the collection of the lesional data during necropsy cannot be excluded, especially at the beginning of the emergence of leptospiral abortions (01/09/14) when the pathologists were less aware of the suspicious lesions. One last limitation is related to the number of cases upon which this study is based to determine the associations between lesions and MAT results. This was mainly due to the epidemic course, with a drop in the icteric cases in October 2014.

Based on overall our findings, an algorithm is proposed to diagnose, in first-line settings, leptospiral infections in cattle in an accurate and economically affordable manner (Figure 2). This approach is particularly reliable in contexts where no vaccination strategies in cattle are in place. In the case of a fetus showing suspicious lesions, MAT would be an initial preferred choice. In two thirds of the cases, it gave a positive result above 1/300 that can likely assign the abortion to an infection of the dam by *Leptospira* spp. PCR on fetal organs was not as efficient for attesting *Leptospira* as a causal agent; only about one quarter of the fetuses showed a positive result. However, using PCR in parallel to MAT allows the confirmation of more cases and the targeting of samples on which culture or molecular typing can be done. In the case of a doubtful diagnosis, complementary tests should be considered, i.e., a PCR on a dam's vaginal discharge/urine [53,54] or a paired sampling of the dam's serum to ascertain an increase in the serological titer [49]. This laboratory approach—knowing the history of vaccination, reproduction and clinical symptoms in the dams—enables the accurate etiological diagnosis of bovine leptospirosis caused by non-maintenance *Leptospira* spp.

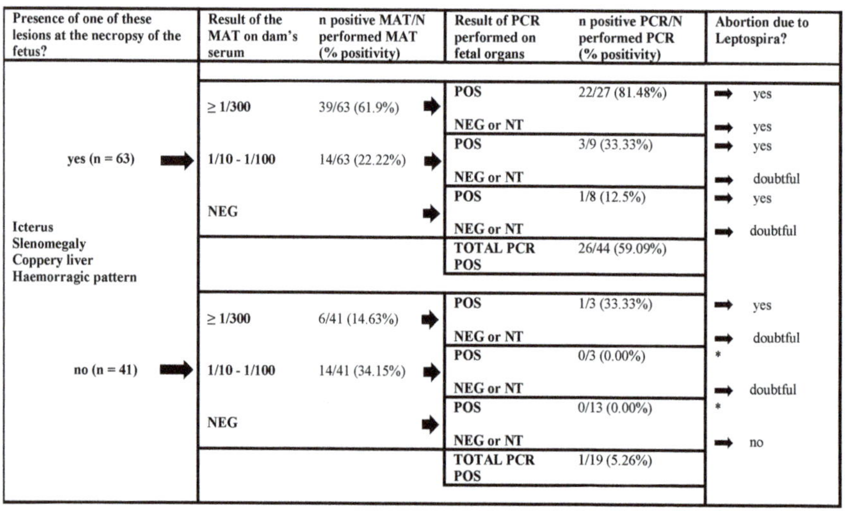

Figure 2. Recommended diagnostic approach for abortuses according to the results of necropsy, MAT on a dam's serum and PCR on fetal organs performed in series, for a conclusion about *Leptospira* spp. as the cause of the abortion. NT: non tested, * did not occur in this study.

4. Materials and Methods

4.1. Study Protocol

The bovine abortuses selected for this study (116) were declared in southern Belgium (Walloon Region) between 27 August 2014 and 15 June 2015 and sampled within the framework of a national surveillance program of abortions in cattle. The study period relates to a time where unusual manifestations of congenital jaundice cases in bovine fetuses were firstly observed in Belgium [30]. In order to have equal representations of icteric and non-icteric abortuses, non-icteric cases were randomly selected among those declared in the same timeframe. The herds where these abortions occurred were not vaccinated as no *Leptospira* vaccine has marketing authorization in Belgium.

4.2. Necropsy and Sample Collection

All abortions were subjected to a standardized necropsy. The thoracic, abdominal and pelvic cavities were examined, as were the brain, the content of the abomasum and—if present—the placental cotyledons. Any lesion was systematically recorded, including (1) icterus, corresponding to a yellow coloring of the connective tissues, especially the visceral adipose ones; (2) splenomegaly, defined as an enlargement of the spleen with a muddy consistency of the organ; (3) coppery liver, the yellow coloration of the hepatic parenchyma with the absence of the modification of the liver volume or its consistency; (4) peri-renal hemorrhages, bleeding under the renal capsule; and (5) generalized hemorrhagic edema (subcutaneous and cavitary) associated, in some cases, with multifocal hemorrhages at the surfaces of organs. The estimation of the gestational ages of the abortuses was made based on the fetal length and external characteristics [55]. Pleural fluid from abortuses (n = 26) was sampled during necropsy, and serum from the dams (n = 108) was collected around the time of the abortion. Samples were stored at 4 °C or at −20 °C for subsequent processing if the tests could not be performed in the immediate days after necropsy.

4.3. Strain and Culture Conditions

Leptospira strains used for the MAT serological test were maintained in liquid Ellinghausen-McCullough-Johnson-Harris (EMJH) medium supplemented with 0.2% w/v yeast extract and 10% fetal calf serum (PAA Laboratories GmbH). The cultures were grown at 29 °C and inoculated weekly by 1:50 dilution. Strains were quality controlled twice-yearly against a panel of hyperimmune positive serovar-specific sera (AMC, Amsterdam).

4.4. MAT

MAT was performed on fetal pleural fluids and the dams' sera as recommended by the OIE manual [49]. The *Leptospira* strains used in the MAT analysis of this study included pathogenic isolates belonging to twelve different serogroups as previously described [34]. Briefly, the sera were diluted in PBS and incubated with live strains at the desired threshold dilution in single well of a 96-well plate. The agglutination reaction was achieved following a 1 h incubation at 37 °C. A loopful of the reaction mix was placed on a slide, and a reading acquired under an upright microscope at 20x magnification (Olympus, Japan). Serum was considered positive if it presented agglutination in at least one of the tested serogroups. Endpoint titers were determined by starting from an initial dilution of 1:10 and using a three-fold dilution until the last condition showing 50% agglutination. The presumptive serogroup was identified by taking the highest agglutination titer of the serum with one particular pathogenic *Leptospira* strain.

4.5. DNA Extraction and Diagnostic Real Time PCR

For nucleic acid extraction, available tissue samples (spleen, placenta, kidneys, adrenal glands, liver, lung, brain or hepatic lymph nodes) were macerated in physiological water, and 200 µL of this homogenate was mixed with 235 μL of lysis-binding solution (MagMax, Applied Biosystem, Paisley, UK) and supplemented with lysozyme at a final concentration of 1 mg/mL. After incubation for one hour at 37 °C, samples were processed for extraction as defined by the manufacturer (MagMax nucleic acid extraction kit, Applied Biosystem). Alternatively, about 20 mg of the roughly chiseled organ was mixed with 180 μL of solution NM1 (LSI MagVet™ Universal Isolation Kit, Thermo Fisher Scientific, Paisley, UK) and, after 1 min of agitation, incubated for 30 min at 70 °C +/− 5 °C. The lysate was then agitated and centrifuged for 1 min at 6000 g. The DNA purification was performed with an automat KingFisher Flex 96™ according to manufacturer's instructions. The diagnostic PCR, targeting the *Lipl32*gene, was performed with in house primers [56] or with the LSI TaqVet™ PathoLept™ (Thermo Fisher Scientific, Merelbeke, BE) as defined by the manufacturer. For in house settings, the reaction mix consisted of 2x absolute mix (Applied Biosystem), 300 μM each of forward and reverse primers and 100 μM of the TaqMan probe. The thermal conditions were those defined elsewhere [56]. The PCR reactions were performed in triplicate on an ABI7500 thermocycler (Thermo Fisher Scientific). Samples showing amplification in at least one PCR reaction (Ct < 45) were considered as positive.

4.6. High-Resolution Melting Analysis

The molecular typing of positive diagnostic PCR samples was based on the polymorphism of the fibronectin-binding protein gene (*lfb1*) [51,52]. This method has been proven as reliable to distinguish genomic *Leptospira* species and to investigate outbreaks, because it does not require samples to have as high bacterial loads as other methods. Polymorphism analysis was achieved with both high-resolution melting analysis (HRMA) and, if PCR amplicons were sufficiently loaded, by *lfb1* gene fragment sequencing. PCR amplifications were achieved using primers previously described [51,52] on a Light Cycler 480 II (Roche) using the LightCycler FastStart DNA Master SYBR Green I (Roche Applied Science). The cycling conditions included a holding step at 95 °C for 10 min; a real-time run of 95 °C for 8 s, 60 °C for 5 s, and 72 °C for 12 s for 45 cycles; and a melting step of 30 °C (from 65 °C to 95 °C) with an acquisition rate of 5 °C. Melting curve analyses were performed with the Light Cycler 480 software release 1.5.1.62 by using default settings.

4.7. Sequencing

PCR amplicons were collected and sent to GENEWIZ (United Kingdom) for outsourcing the sequencing. The subcontracted service included PCR amplicon purification and sequencing by the dideoxy chain-termination procedure. The quality of the chromatograms was analyzed with the Bionumerics software package V 6.6 (Applied Maths, Belgium). An assembly between forward and reverse sequences was obtained with default settings. The sequences obtained in this study have been deposited in GenBank (accession numbers KY373222 to KY373229). Multiple alignments were done, setting the gap penalty at 0%, and cluster analysis was performed using the unweighted-pair group method with an arithmetic mean algorithm (UPGMA).

4.8. Statistical Analysis

Associations between necropsy findings and MAT/PCR results were evaluated with the Chi-square test, or Fisher's exact test in cases of any expected frequency inferior to 5 (significance level of $p < 0.05$). The statistical tests were performed with SigmaPlot 11.0 (Systat Software, San Jose, CA). In addition, for the analysis of the lesional data, odds ratios were calculated with 95% confidence intervals (95% CI). For the agreement between the diagnostic tests (MAT and PCR), Cohen's kappa statistic with 95% confidence intervals was calculated. The interpretation for the kappa statistic was as follows: <0.2, slight agreement; 0.2–0.4, fair agreement; 0.4–0.6, moderate agreement; 0.6–0.8, substantial agreement; and >0.8, almost perfect agreement [57]. The kappa coefficients were calculated using WIN EPISCOPE 2.0 [58]. In addition to the kappa coefficients, the relative sensitivities and specificities of the MAT towards PCR were calculated with 95% "exact" Clopper–Pearson confidence intervals.

Supplementary Materials: The following are available online at http://www.mdpi.com/2076-0817/9/6/413/s1, Table S1. Detailed results of necropsy, the MAT and the real-time PCR performed on 116 abortuses for the diagnosis of leptospirosis. SPL: splenomegaly; CL: coppery liver; PRH: peri-renal hemorrhages; SD: serum of the dam; PLF: pleural fluid. Dilution*: agglutination less than 50%. Table S2: Cross-reactions between serovars of *Leptospira* spp. observed on 75 dams' sera submitted to MAT (cut-off of 1/10). Table S3: Comparison of results of PCR performed on fetuses and results of MAT performed on the serum of the corresponding dams. Figure S1: Melting curves derived from analysis of the *lfb1* sequence polymorphisms in *Leptospira* spp. icteric abortions by HRMA.

Author Contributions: Conceptualization, F.G. and M.M.; Data curation, F.G., T.P., S.B., L.D., M.S. and M.M.; Formal analysis, F.G., R.B., S.B., L.D. and M.M.; Investigation, F.G., R.B., T.P., S.B., L.D., D.F., M.S. and M.M.; Resources, F.G., R.B., T.P., S.B., L.D., M.S. and M.M.; Supervision, F.G. and M.M.; Writing—original draft, F.G. and M.M.; Writing—review & editing, F.G., R.B., T.P., S.B., L.D., D.F., M.S. and M.M. All authors have read and agree to the published version of the manuscript.

Funding: This research received no external funding.

Acknowledgments: The authors thank Behaeghel I., Forrez E., Charrot J., Evrard J. and Quinet C. for their collaborative support. Gratitude is also expressed to Marin M., Vannoorenberghe P. and the rest of the technical laboratory staff of the Veterinary Bacteriology and to Cuvelier P. from the laboratory staff of ARSIA for the technical help.

Conflicts of Interest: The authors declare no conflict of interest.

Ethics statement: This study is a retrospective investigation of samples collected by the first line and the reference laboratory during the national surveillance program of abortions in cattle for brucellosis and now characterized for leptospiral infection.

References

1. Vincent, A.T.; Schiettekatte, O.; Goarant, C.; Neela, V.K.; Bernet, E.; Thibeaux, R.; Ismail, N.; Khalid, M.K.N.M.; Amran, F.; Masuzawa, T.; et al. Revisiting the taxonomy and evolution of pathogenicity of the genus Leptospira through the prism of genomics. *PLoS Negl. Trop. Dis.* **2019**, *13*, e0007270. [CrossRef] [PubMed]
2. Picardeau, M. Diagnosis and epidemiology of leptospirosis. *Med. Mal. Infect.* **2013**, *43*, 1–9. [CrossRef] [PubMed]
3. Lehmann, J.S.; Matthias, M.A.; Vinetz, J.M.; Fouts, D.E. Leptospiral pathogenomics. *Pathogens* **2014**, *3*, 280–308. [CrossRef]
4. Ellis, W.A.; Michna, S.W. Bovine leptospirosis: A serological and clinical study. *Vet. Rec.* **1976**, *99*, 387–391. [CrossRef] [PubMed]
5. Hassanpour, A.; Mousavi, G.H. A case report of leptospira grippotyphosa in the Azerbaijan buffalo in Iran. *Ital. J. Anim. Sci.* **2007**, *6*, 893–895. [CrossRef]
6. Bahari, A.; Abdollahpour, G.; Sadeghi-Nasab, A.; Sattari Tabrizi, S.; Yavari, M.; Dadmehr, B. A serological survey on leptospirosis in aborted dairy cattle in industrial farms of Hamedan suburb, Iran. *Iran. J. Vet. Res.* **2011**, *12*, 337–339.
7. Marquez, A.; Ulivieri, T.; Benoit, E.; Kodjo, A.; Lattard, V. House mice as a real sanitary threat of human and animal leptospirosis: Proposal for integrated management. *Biomed Res. Int.* **2019**, *2019*, 3794876. [CrossRef]
8. Ayral, F.; Zilber, A.L.; Bicout, D.J.; Kodjo, A.; Artois, M.; Djelouadji, Z. Distribution of leptospira interrogans by multispacer sequence typing in urban Norway rats (Rattus norvegicus): A survey in France in 2011–2013. *PLoS ONE* **2015**, *10*, e0139604. [CrossRef]
9. Hanson, L.E. Bovine leptospirosis. *J. Dairy Sci.* **1976**, *59*, 1166–1170. [CrossRef]
10. Adler, B.; de la Pena Moctezuma, A. Leptospira and leptospirosis. *Vet. Microbiol.* **2010**, *140*, 287–296. [CrossRef]
11. Cordy, D.R.; Jasper, D.E. The pathology of an acute hemolytic anemia of cattle in California associated with Leptospira. *J. Am. Vet. Med. Assoc.* **1952**, *120*, 175–178. [PubMed]
12. Hathaway, S.C.; Little, T.W.; Jones, T.W.; Stevens, H.; Butland, R.W. Infection by leptospires of the Pomona serogroup in cattle and pigs in south west England. *Vet. Rec.* **1984**, *115*, 246–248. [CrossRef] [PubMed]
13. Pritchard, G.C.; Borland, E.D.; Wood, L.; Pritchard, D.G. Severe disease in a dairy herd associated with acute infection with bovine virus diarrhoea virus, Leptospira harjo and Coxiella burnetii. *Vet. Rec.* **1989**, *124*, 625–629. [CrossRef] [PubMed]

14. Gummow, B.; Myburgh, J.G.; Thompson, P.N.; van der Lugt, J.J.; Spencer, B.T. Three case studies involving Leptospira interrogans serovar pomona infection in mixed farming units. *J. S. Afr. Vet. Assoc.* **1999**, *70*, 29–34. [CrossRef] [PubMed]
15. Smith, L.L. Chronic Leptospira hardjo and Leptospira hebdomadis infection in a fifty cow herd of dairy cattle, a case history. *Proc. Annu. Meet. U. S. Anim. Heal. Assoc.* **1969**, *73*, 181–183.
16. Ellis, W.A.; O'Brien, J.J.; Bryson, D.G.; Mackie, D.P. Bovine leptospirosis: Some clinical features of serovar hardjo infection. *Vet. Rec.* **1985**, *117*, 101–104. [CrossRef]
17. Ellis, W.A.; O'Brien, J.J.; Cassells, J.A.; Neill, S.D.; Hanna, J. Excretion of Leptospira interrogans serovar hardjo following calving or abortion. *Res. Vet. Sci.* **1985**, *39*, 296–298. [CrossRef]
18. Ellis, W.A.; O'Brien, J.J.; Cassells, J. Role of cattle in the maintenance of Leptospira interrogans serotype hardjo infection in Northern Ireland. *Vet. Rec.* **1981**, *108*, 555–557. [CrossRef]
19. Dhaliwal, G.S.; Murray, R.D.; Dobson, H.; Ellis, W.A. Effect of Leptospira interrogans serovar hardjo infection on progesterone concentrations in heifers. *Vet. Rec.* **1997**, *140*, 19–20. [CrossRef]
20. Dhaliwal, G.S.; Murray, R.D.; Dobson, H.; Montgomery, J.; Ellis, W.A. Effect of Leptospira interrogans serovar hardjo infection on milk yield in endemically infected dairy herds. *Vet. Rec.* **1996**, *139*, 319–320. [CrossRef]
21. Otaka, D.Y.; Martins, G.; Hamond, C.; Penna, B.; Medeiros, M.A.; Lilenbaum, W. Serology and PCR for bovine leptospirosis: Herd and individual approaches. *Vet. Rec.* **2012**, *170*, 338. [CrossRef] [PubMed]
22. Smyth, J.A.; Fitzpatrick, D.A.; Ellis, W.A. Stillbirth/perinatal weak calf syndrome: A study of calves infected with Leptospira. *Vet. Rec.* **1999**, *145*, 539–542. [CrossRef] [PubMed]
23. Lilenbaum, W.; Souza, G.N. Factors associated with bovine leptospirosis in Rio de Janeiro, Brazil. *Res. Vet. Sci.* **2003**, *75*, 249–251. [CrossRef]
24. Dhaliwal, G.S.; Murray, R.D.; Ellis, W.A. Reproductive performance of dairy herds infected with Leptospira interrogans serovar hardjo relative to the year of diagnosis. *Vet. Rec.* **1996**, *138*, 272–276. [CrossRef] [PubMed]
25. Langoni, H.; De Souza, L.C.; Da Silva, A.V.; Luvizotto, M.C.R.; Paes, A.C.; Lucheis, S.B. Incidence of leptospiral abortion in Brazilian dairy cattle. *Prev. Vet. Med.* **1999**, *40*, 271–275. [CrossRef]
26. Escamilla, H.P.; Martínez, M.J.J.; Medina, C.M.; Morales, S.E. Frequency and causes of infectious abortion in a dairy herd in Queretaro, Mexico. *Can. J. Vet. Res.* **2007**, *71*, 314–317. [PubMed]
27. Lucchese, L.; Benkirane, A.; Hakimi, I.; Idrissi, A. El Seroprevalence study of the main causes of abortion in dairy cattle in Morocco. *Vet. Ital.* **2016**, *52*, 13–19. [PubMed]
28. Bahaman, A.R.; Ibrahim, A.L.; Stallman, N.D.; Tinniswood, R.D. The bacteriological prevalence of leptospiral infection in cattle and buffaloes in West Malaysia. *Epidemiol. Infect.* **1988**, *100*, 239–246. [CrossRef] [PubMed]
29. Feresu, S.B.; Bolin, C.A.; Korver, H.; Van de Kemp, H. Identification of leptospires of the Pomona and Grippotyphosa serogroups isolated from cattle in Zimbabwe. *Res. Vet. Sci.* **1995**, *59*, 92–94. [CrossRef]
30. Abdollahpour, G.; English, A.W.; Tasler, J. Isolation of Leptospira interrogans serovar grippotyphosa from a heifer in New South Wales. *Aust. Vet. J.* **1996**, *73*, 109–110. [CrossRef]
31. Ellis, W.A. Animal leptospirosis. In *Leptospira and Leptospirosis*; Adler, B., Ed.; Springer: Berlin/Heidelberg, Germany, 2015; pp. 99–137. ISBN 3-662-45058-5.
32. Delooz, L.; Mori, M.; Petitjean, T.; Evrard, J.; Czaplicki, G.; Saegerman, C. Congenital jaundice in bovine aborted foetuses: An emerging syndrome in Southern Belgium. *Transbound. Emerg. Dis.* **2015**, *62*, 124–126. [CrossRef] [PubMed]
33. Delooz, L.; Czaplicki, G.; Gregoire, F.; Dal Pozzo, F.; Pez, F.; Kodjo, A.; Saegerman, C. Serogroups and genotypes of Leptospira spp. strains from bovine aborted foetuses. *Transbound. Emerg. Dis.* **2018**, *65*, 158–165. [CrossRef] [PubMed]
34. Mori, M.; Van Esbroeck, M.; Depoorter, S.; Decaluwe, W.; Vandecasteele, S.J.; Fretin, D.; Reynders, M. Outbreak of leptospirosis during a scout camp in the Luxembourg Belgian province, Belgium, summer 2012. *Epidemiol. Infect.* **2015**, *143*, 1761–1766. [CrossRef] [PubMed]
35. Ellis, W.A.; Michna, S.W. Experimental leptospiral abortion in cattle. *Vet. Rec.* **1974**, *94*, 255. [CrossRef] [PubMed]
36. Delooz, L.; Saegerman, C.; Quinet, C.; Petitjean, T.; De Regge, N.; Cay, B. Resurgence of schmallenberg virus in belgium after 3 years of epidemiological silence. *Transbound. Emerg. Dis.* **2016**, *64*, 1641–1642. [CrossRef]
37. Comité Scientifique de l'Agence Fédérale pour la Sécurité de la Chaîne Alimentaire. *AFSCA Réémergence de la brucellose bovine en Belgique entre 2010 et 2013 (SciCom N° 2011/10)*; AFSCA: Brussels, Belgium, 2016.

38. Picardeau, M. Virulence of the zoonotic agent of leptospirosis: Still terra incognita? *Nat. Rev. Microbiol.* **2017**, *15*, 297. [CrossRef]
39. Pijnacker, R.; Goris, M.G.; te Wierik, M.J.; Broens, E.M.; van der Giessen, J.W.; de Rosa, M.; Wagenaar, J.A.; Hartskeerl, R.A.; Notermans, D.W.; Maassen, K.; et al. Marked increase in leptospirosis infections in humans and dogs in the Netherlands, 2014. *Eurosurveillance* **2016**, *21*. [CrossRef]
40. Institut de Santé Publique (WIV-ISP). *Zoonoses et maladies à transmission vectorielles—Surveillance épidémiologique en Belgique, 2013 et 2014*; Institut de Santé Publique: Ixelles, Belgium, 2015.
41. Ayral, F.C.; Bicout, D.J.; Pereira, H.; Artois, M.; Kodjo, A. Distribution of Leptospira serogroups in cattle herds and dogs in France. *Am. J. Trop. Med. Hyg.* **2014**, *91*, 756–759. [CrossRef]
42. Tagliabue, S.; Figarolli, B.M.; D'Incau, M.; Foschi, G.; Gennero, M.S.; Giordani, R.; Natale, A.; Papa, P.; Ponti, N.; Scaltrito, D.; et al. Serological surveillance of Leptospirosis in Italy: Two year national data (2010 2011). *Vet. Ital.* **2016**, *52*, 129–138.
43. Alonso-Andicoberry, C.; García-Peña, F.J.; Pereira-Bueno, J.; Costas, E.; Ortega-Mora, L.M. Herd-level risk factors associated with Leptospira spp. seroprevalence in dairy and beef cattle in Spain. *Prev. Vet. Med.* **2001**, *52*, 109–117. [CrossRef]
44. Atxaerandio, R.; Aduriz, G.; Ziluaga, I.; Esteban, J.I.; Maranda, L.; Mainar-Jaime, R.C. Serological evidence of Leptospira interrogans serovar Bratislava infection and its association with abortions in cattle in northern Spain. *Vet. Rec.* **2005**, *156*, 376. [CrossRef] [PubMed]
45. Behaeghel, I.; Butaye, P.; Goossens, E. Evolution of leptospirosis in Belgian dogs from 2002 to 2009. In Proceedings of the 54th BSAVA Annual Congress, Birmingham, UK, 31 March–3 April 2011.
46. Desmecht, M.; Colin, G. Isolation in Belgium of Leptospira grippotyphosa from a muskrat. *Ann. Med. Vet.* **1988**, *132*, 693–696.
47. Lernout, T.; Van Esbroeck, M. Surveillance Épidémiologique de la Leptospirose Leptospira spp.—2018. Available online: https://epidemio.wiv-isp.be/ID/diseases/Documents/Reports2018/Lepto_2018_fr.pdf (accessed on 9 April 2020).
48. Goris, M.G.A.; Leeflang, M.M.G.; Boer, K.R.; Goeijenbier, M.; van Gorp, E.C.M.; Wagenaar, J.F.P.; Hartskeerl, R.A. Establishment of valid laboratory case definition for human leptospirosis. *J. Bacteriol. Parasitol.* **2011**, *3*. [CrossRef]
49. World Organisation for Animal Health (OIE). Chapter 2.1.9. Leptospirosis. In *Manual for Diagnosis Tests and Vaccines for Terrestrial Animals*; World Organisation for Animal Health: Paris, France, 2014.
50. Dhaliwal, G.S.; Murray, R.D.; Dobson, H.; Montgomery, J.; Ellis, W.A.; Baker, J.R. Presence of antigen and antibodies in serum and genital discharges of heifers after experimental intrauterine inoculation with Leptospira interrogans serovar hardjo. *Res. Vet. Sci.* **1996**, *60*, 157–162. [CrossRef]
51. Perez, J.; Goarant, C. Rapid Leptospira identification by direct sequencing of the diagnostic PCR products in New Caledonia. *Bmc Microbiol.* **2010**, *10*, 325. [CrossRef] [PubMed]
52. Naze, F.; Desvars, A.; Picardeau, M.; Bourhy, P.; Michault, A. Use of a new high resolution melting method for genotyping pathogenic leptospira spp. *PLoS ONE* **2015**, *10*, e0127430. [CrossRef]
53. Loureiro, A.P.; Pestana, C.; Medeiros, M.A.; Lilenbaum, W. High frequency of leptospiral vaginal carriers among slaughtered cows. *Anim. Reprod. Sci.* **2017**, *178*, 50e4. [CrossRef]
54. Pinna, A.; Martins, G.; Loureiro, A.P.; Lilenbaum, W. Detection of bovine carriers of leptospira by serological, bacteriological, and molecular tools. *Trop. Anim. Health Prod.* **2018**, *50*, 883–888. [CrossRef]
55. Njaa, B.L. *Kirkbride's Diagnosis of Abortion and Neonatal Loss in Animals*, 4th ed.; Njaa, B.L., Ed.; Wiley-Blackwell: West Sussex, UK, 2012.
56. Stoddard, R.A.; Gee, J.E.; Wilkins, P.P.; McCaustland, K.; Hoffmaster, A.R. Detection of pathogenic Leptospira spp. through TaqMan polymerase chain reaction targeting the LipL32 gene. *Diagn. Microbiol. Infect. Dis.* **2009**, *64*, 247–255. [CrossRef]
57. Petrie, A.; Watson, P. *Statistics for Veterinary and Animal Science*; Blackwell Science: London, UK, 1999.
58. Thrusfield, M.; Ortega, C.; de Blas, I.; Noordhuizen, J.P.; Frankena, K. WIN EPISCOPE 2.0: Improved epidemiological software for veterinary medicine. *Vet. Rec.* **2001**, *148*, 567–572. [CrossRef]

© 2020 by the authors. Licensee MDPI, Basel, Switzerland. This article is an open access article distributed under the terms and conditions of the Creative Commons Attribution (CC BY) license (http://creativecommons.org/licenses/by/4.0/).

pathogens

MDPI

Article

Serological Survey and Molecular Typing Reveal New *Leptospira* Serogroup Pomona Strains among Pigs of Northern Italy

Cristina Bertasio [1,*], Alice Papetti [1], Erika Scaltriti [2], Silvia Tagliabue [1], Mario D'Incau [1] and Maria Beatrice Boniotti [1]

1 National Reference Centre for Animal Leptospirosis (NRCL), Istituto Zooprofilattico Sperimentale della Lombardia e dell' Emilia Romagna "Bruno Ubertini", via Bianchi 7/9, 25121 Brescia, Italy; alice.papetti@izsler.it (A.P.); tagliabues@gmail.com (S.T.); mario.dincau@izsler.it (M.D.); mariabeatrice.boniotti@izsler.it (M.B.B.)

2 Risk Analysis and Genomic Epidemiology Unit, Istituto Zooprofilattico Sperimentale della Lombardia e dell' Emilia Romagna "Bruno Ubertini", Strada dei Mercati 13/a, 43126 Parma, Italy; erika.scaltriti@izsler.it

* Correspondence: cristina.bertasio@izsler.it; Tel.: +030-22-90-309

Received: 2 April 2020; Accepted: 27 April 2020; Published: 29 April 2020

Abstract: Swine act as both maintenance and incidental hosts of pathogenic *Leptospira* spp. Here, a serological test was performed on 131,660 pig sera collected between 2002 and 2017 from 4715 farms in Northern Italy. A positivity rate of 13.05% was determined. Australis was the most frequently identified serogroup (77.29%), followed by Pomona (18.47%), Tarassovi (1.51%) and Icterohaemorrhagie (1.40%). Culture isolation and real-time Polymerase chain reaction (PCR) were carried out on 347 kidneys and 470 clinical samples, respectively. Overall, 133 strains were cultured successfully and 43 randomly chosen isolates were identified as serogroup Pomona. Multi-locus sequence typing (MLST) revealed that 41 isolates and 8 DNA extracted from biological samples belonged to sequence type 140. Using a multiple-locus, variable-number tandem repeat analysis, 43 samples produced identical profiles but, after 2014, three new *Leptospira interrogans* serogroup Pomona genotypes were observed. Interestingly, two isolates showed new MLST profiles and an unclassified identification by monoclonal antibodies. The 16S rRNA gene sequencing clustered them into *L. kirschneri* species and a core genome MLST analysis revealed an allelic identity of 96% compared with Mozdok strains. Genotyping allowed us to discriminate leptospires and to identify new emerging strains. The accurate identification of infective strains is required for formulating preventive methods and intervention strategies.

Keywords: Leptospirosis; pig; MAT; real-time PCR; genotyping; epidemiology

1. Introduction

Leptospirosis is the most widespread zoonosis worldwide and it is caused by an infection with any of the pathogenic members of the genus *Leptospira*. While, in theory, any pathogenic *Leptospira* may infect any animal species, leptospirosis is a disease that shows a natural nidality, and each serovar tends to be maintained in specific maintenance hosts. For example, serovar Bratislava is associated with pigs and horses, Canicola is commonly found in dogs, Icterohaemorrhagiae in rats, Pomona in pigs and Hardjo in cattle [1–3]. Leptospires persist in the kidneys of carrier animals and are excreted in urine and genital fluids [3–6]. In this way, they are spread through the environment where they can survive for long periods, contaminating the surface water, soil and muddy areas [2–5,7]. Chronically infected animals may remain carriers over years and act as reservoirs for the infection of other animals and humans [3,5]. Despite rat being the main reservoir for these bacteria, many other mammals can act as carriers. Swine, for example, act as a maintenance host for leptospires and are a possible source of human and domestic animal infections [3,5]. Historically, pigs act as a maintenance host for the serovars Bratislava, Pomona and Tarassovi, while among the incidental serovars, the most important in pigs are those belonging to the Icterohaemorrhagiae, Canicola and Grippotyphosa serogroups [3–5].

In Italy, swine act commonly as carriers for serovar Bratislava, belonging to the Australis serogroup, and serovar Pomona, belonging to the Pomona serogroup. Additionally, serovar Tarassovi causes commonly incidental infections [8]. Endemic infections in swine herds generally remain subclinical, and the only clinical symptoms are reproductive disorders, such as late-term abortions, and increases in mummified, stillborn and weak piglets. However, leptospires can also cause severe diseases depending mainly on the infecting serovar and the age of the animal [4]. Once the infective agent has entered a farm, it spreads very easily, mostly among fattening pigs, both through direct (contaminated urine) and indirect (infected feed, water and environment) contact [8]. Vaccination, therapy and farm management can all be used to limit the spread of infection in a breeding herd.

Until 2010, in Italy, a trivalent vaccine against the serogroups Australis, Pomona and Tarassovi was available for swine. Nevertheless, very few breeders have adopted vaccination practices, primarily because the risk of leptospirosis is poorly understood by farmers and because the treatments and/or control strategies for other more virulent diseases are more important to the farm's economy. In 2011, vaccinations were completely abandoned because no commercial vaccine was available for pigs. Starting from that moment, the main infection control strategy was the management of the breeding herd through the prevention of direct or indirect contact with free-living vectors or other domestic stock. Strong monitoring and surveillance systems are needed to better understand the disease epidemiology, and strict biosecurity tools should be applied to limit the transmission of the infective agent. Leptospirosis is not included in the Office International des Epizooties (OIE) list of notifiable terrestrial and aquatic animal diseases; however, in Italy, it is considered, by the current rules, a notifiable infection [9]. In cases of clinical suspicion, which are confirmed by a serological examination, the outbreak must be officially reported to the authorities. The farm is seized and sanitary measures, aimed at the eradication of the infective agent, are implemented. The interruption of transmission through the isolation of the infected animals is the first action that must be implemented. Only when all the animals become seronegative may the infection be considered eradicated, and the restrictive measures can be removed [9,10]. This can take months, or even years, resulting in significant economic losses.

In the laboratory, serological testing is the most widely used means for diagnosing leptospirosis, and the Microscopic Agglutination Test (MAT) is the serological gold standard method reported by the OIE [11]. Ideally, antigens selected for the MAT should include representative strains of all serogroups known to circulate in the study region as well as those known to be maintained elsewhere by the host species being analyzed.

The OIE indicates that strain isolation should be attempted from biological samples. Cultured strains can be studied in depth, and the serogroup and serovar status should be determined using polyclonal and Monoclonal Antibodies (mAbs), respectively [11]. Nevertheless, isolation does not contribute directly to acute diagnosis of leptospirosis because *Leptospira* grows slowly over weeks [12]. With the advent of molecular methods, the diagnosis of a *Leptospira* infection has become simpler and faster. Over the years, a number of real-time PCR-based methods have been described for the direct identification of *Leptospira* nucleic acids in biological samples, both for detection of pathogenic and environmental species [13–20]. Real-time PCR provides advantages over the classical conventional reference methods used to diagnose leptospirosis (such as MAT and isolation) including reduced turnaround and hands-on times, low carryover contamination risks and higher sensitivity and specificity levels.

Furthermore, in the last few years, genotypic classification has increasingly integrated traditional serological classification in a wide variety of bacterial species. For *Leptospira*, the classical classification into serogroups and serovars has been joined by molecular typing. Several molecular techniques, such as Multiple-Locus Variable-Number Tandem-Repeat (VNTR) Analysis (MLVA) and Multi-Locus Sequence Typing (MLST), aimed at identifying bacteria by examining individual genomic profiles, have been used to investigate the epidemiology of *Leptospira*. The MLVA method is a useful typing tool for identifying *Leptospira* genotypes by providing information on genetic relationships among isolates for the surveillance of *Leptospira* populations. Salaün and colleagues [21] developed an MLVA method that permits the discrimination of the three most common pathogenic species of *Leptospira* (*L. interrogans*, *L. kirschneri* and *L. borgpetersenii*) through the analysis of five loci of repetitive unit sequences (VNTR 4, 7, 10, Lb4 and Lb5).

In 2013, Boonsilp and colleagues [22] proposed an MLST scheme based on the sequencing of seven housekeeping genes. This protocol is optimized to work on isolate samples, but it is not effective on biological samples, owing to the poor bacterial load. Thanks to a new protocol published in 2016 by Weiss [23], it is now possible to perform MLST directly on DNA extracted from biological samples, overcoming the long and difficult isolation step. With Whole-Genome Sequencing (WGS), new genome-based analyses have been developed for bacterial typing. Very recently Guglielmini et al. [24] published a core genome MLST (cgMLST) scheme based on 545 highly conserved genes in *Leptospira*, increasing the discriminatory power necessary to distinguish isolates.

This study aims to provide a serological survey of the prevalence of *Leptospira* in pigs in Northern Italy and to characterize the circulating strains using innovative molecular techniques. Our goal is to understand the epidemiology of swine leptospirosis in this specific area and to identify possible new emerging strains to address preventive and control measures and to reduce the risk of infection in swine herds.

2. Results

2.1. Microscopic Agglutination Test (MAT)

Using MAT, 17,184 out of 131,660 sera collected from 4715 farms were identified as positive, with a seropositivity of 13.05% (cut-off ≥ 1:100). The minimum percentage of positive sera was observed in 2006 (8.46%), the maximum one in 2012 (18.31%). Among the positive samples, 13,809 (10.49% of the total samples and 80.36% of the positive samples) tested positive for one serogroup, while 3375 (2.56% of the total samples and 19.64% of the positive samples) were positive for more than one serogroup. Antibody titers against the serogroup Australis occurred in 8.11% of the total samples and 77.29% of the single positive samples, followed by Pomona (1.94% and 18.47%, respectively), Tarassovi (0.16% and 1.51%, respectively), Icterohaemorrhagiae (0.15% and 1.40%, respectively), Sejroe (0.07% and 0.67%, respectively), Ballum (0.03% and 0.32%, respectively), Canicola (0.02% and 0.22%, respectively) and Grippotyphosa (0.01% and 0.12%, respectively) (Table 1). Using the cut-off ≥ 1:200, the Australis-positive samples decreased to 66.75%, the Pomona-positive samples increased to 29.28%, and the other serogroups maintained similar values (data not shown).

For positivity to more than one serogroup (Table 2), the predominant combination was represented by Australis-Pomona, having a prevalence of 54.72% (1847 out of 3375 multiple-positive sera) followed by Australis-Icterohaemorrhagiae-Pomona with 16.91% (571/3375), Australis-Tarassovi and Australis-Icterohaemorrhagiae (6.40% and 6.34%, respectively). The Icterohaemorrhagiae-Pomona combination was found in 2.75% of samples, while other combinations were revealed at levels < 2%. Australis was present in 3156 (93.51%), while Pomona was present in 2855 (84.59%), of 3375 multiple-positive sera.

Table 1. Numbers and percentages of swine serum samples that tested positive by Microscopic Agglutination Test (MAT) for Leptospira serogroups from 2002 to 2017. Only positivity to one serogroup was included.

Year	Serogroups								No. of Positive Samples
	A	B	C	G	I	P	S	T	
2002	1078	6	5	1	25	379	4	117	1615
2003	925	2	1	6	22	240	5	34	1235
2004	690	2			28	275	4	16	1015
2005	662	1	1		7	283	5	6	965
2006	301	7	2		9	440	7	5	771
2007	933	3	1	2	27	245	16	7	1234
2008	824	1			1	115	7	9	957
2009	647	5	1		6	157	1	2	819
2010	793		3		5	90	12		903
2011	639	2			6	52	1	1	701
2012	1414	3	8	2	7	64	10	1	1509
2013	334	2			11	35	6	1	389
2014	346	1	1		5	61	2		416
2015	441	5	2	2	6	29	9	3	497
2016	385	3	3		18	70	2	4	485
2017	261	1	3	3	10	16	1	3	298
No. of positive samples	10,673	44	31	16	193	2551	92	209	13,809
Percentage (%) of the positives (N = 13,809)	77.29	0.32	0.22	0.12	1.40	18.47	0.67	1.51	100
Percentage (%) of the total (N = 131,660)	8.11	0.03	0.02	0.01	0.15	1.94	0.07	0.16	10.49

A, Australis; B, Ballum; C, Canicola; G, Grippotyphosa; I, Icterohaemorrhagiae; P, Pomona; S, Sejroe; T, Tarassovi.

The MAT titers of the single positive samples were generally low (1:100 or 1:200) for all the serogroups detected, except for the serogroup Pomona, which showed medium (1:400) or high titers (greater than 1:800) in more than 50% of the samples (Figure 1). In particular, more than 90% of the samples positive for the serogroup Australis showed low antibody titers, while those positive for Pomona showed mainly medium (17.99%) and high (38.79%) titers. Grippotyphosa-positive samples also showed high titers (≥1:800) in more than 35% of the samples (Figure 1). The detailed titer distributions of positive sera reacting to one serogroup are reported in Table 3.

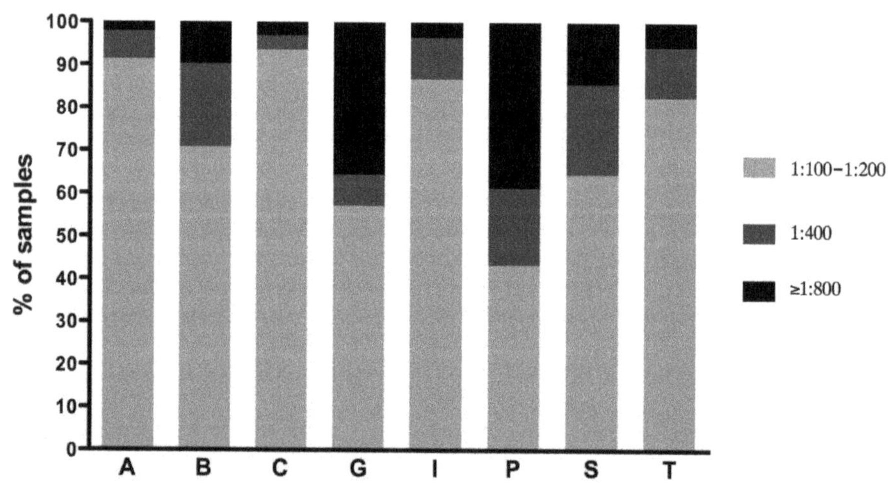

Figure 1. Distribution of MAT titers of samples reacting to one serogroup. Titers were grouped as low (1:100–1:200), medium (1:400) and high (≥1:800). A, Australis; B, Ballum; C, Canicola; G, Grippotyphosa; I, Icterohaemorrhagiae; P, Pomona; S, Sejroe; T, Tarassovi.

Table 2. Sera that react to more than one serogroup by Microscopic Agglutination Test (MAT).

Serogroups Combination	N° of Sera	Serogroups Combination	N° of Sera
A-P	1847	B-P	3
A-I-P	571	C-G-S	3
A-T	216	A-B-S	2
A-I	214	A-C-G	2
I-P	93	A-C-I-P-S	2
A-P-T	66	A-C-S	2
P-T	64	A-G-P-S	2
A-G-P	46	G-S	2
A-G-I-P	44	A-B-C-P	1
A-C-I-P	21	A-B-I-P-T	1
G-P	18	A-C-G-I-P-S-T	1
A-I-P-T	14	A-C-G-S	1
A-S	14	A-C-I-T	1
A-B-I-P	8	A-C-P	1
A-I-P-S	8	A-G-I-S	1
A-I-T	8	A-G-P-T	1
C-I	8	A-I-S	1
A-P-S	7	A-P-S-T	1
C-S	7	A-S-T	1
G-I-P	7	B-G	1
A-B	6	B-I	1
A-B-C-I-P	6	C-G	1
A-G	6	C-G-I	1
P-S	6	C-G-I-S	1
A-B-P	5	C-I-P	1
A-C-I	5	C-P	1
A-G-I	5	C-T	1
A-B-C-G-I-P-S-T	4	G-P-S	1
G-I	4	I-P-T	1
A-B-C	3	I-S	1
A-G-I-P-S	3	I-T	1
N. of total sera: 3375			

A, Australis; B, Ballum; C, Canicola; G, Grippotyphosa; I, Icterohaemorrhagiae; P, Pomona; S, Sejroe; T, Tarassovi.

Table 3. Detailed MAT titers of samples reacting to one serogroup.

Serogroup	Titer							Total
	1:100	1:200	1:400	1:800	1:1600	1:3200	1:6400	
Australis	6901	2836	694	170	43	16	5	10,665
Ballum	14	15	8	3	1	0	0	41
Canicola	22	7	1	0	1	0	0	31
Grippotyphosa	7	1	1	2	0	2	1	14
Icterohaemorragiae	115	29	16	6	0	0	0	166
Pomona	529	516	435	333	255	172	178	2418
Sejroe	35	23	19	6	6	0	1	90
Tarassovi	116	54	24	10	1	1	0	206

Out of 4715 tested farms, a mean of 53.62% resulted as positive for the presence of pathogenic *Leptospira* (Table 4). A farm was considered positive if at least one pig was positive by MAT to one or more *Leptospira* serogroups. The percentage of farms with positive test results ranged between 22.84% (n = 140) in 2006 and 77.52% (n = 231) in 2012.

Table 4. Results of the MAT analysis of tested farms. Number (n) and percentage (%) of positive farms between 2002 and 2017.

Year	Number of Tested Farms	*Leptospira* Positive Farms	
		n	%
2002	452	226	50.00%
2003	455	243	53.41%
2004	528	244	46.21%
2005	478	252	52.72%
2006	613	140	22.84%
2007	332	190	57.23%
2008	265	162	61.13%
2009	246	146	59.35%
2010	225	155	68.89%
2011	202	150	74.26%
2012	298	231	77.52%
2013	135	78	57.78%
2014	130	79	60.77%
2015	128	88	68.75%
2016	128	82	64.06%
2017	101	62	61.39%
Total	4715	2528	53.62%

Outbreaks caused by Australis were stable over the years, with a maximum value of 82.23% in 2012 and a minimum of 66.34% in 2002. This serogroup is responsible for a mean of 75.87% ± 0.15% (Confidence Interval 95%) of the outbreaks during the study period (Figure 2).

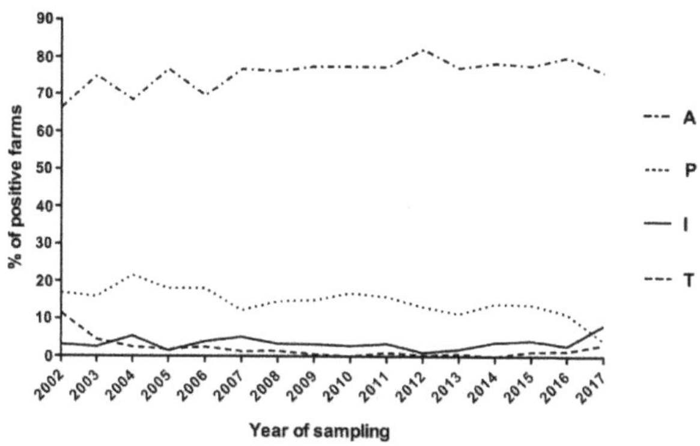

Figure 2. Trend in the main observed outbreaks between 2002 and 2017 (only sera positive for a single serogroup were considered). A, Australis; B, Ballum; C, Canicola; G, Grippotyphosa; I, Icterohaemorrhagiae; P, Pomona; S, Sejroe; T, Tarassovi.

The second serogroup in terms of frequency of outbreak was Pomona, with a mean of 14.56% ± 0.31% (CI 95%) over the 16 years of the study with the maximum value observed in 2004 (21.58%) and the minimum one in 2017 (4.17%). Serogroup Icterohaemorrhagiae was responsible for a mean frequency of 3.52% ± 0.30% (CI 95%), with a peak of 8.33% in 2017. In 2017, an inversion occurred between the Pomona and Icterohaemorrhagiae serogroups: the first decreased to the minimum value (4.17%), while the latter increased to a new maximum level (8.33%). The mean frequency of Tarassovi was 2.10% ± 0.48% (CI 95%) but interestingly, its trend started from 11.49% in 2002 and became null in 2014. For Sejroe, a very low frequency was registered over the years, with values between 0% in 2014 and 2.84% in 2016 (mean of 1.64% ± 0.19%, CI 95%). For the other serogroups, the mean prevalence was less than 2% with rare exceptions during the considered time (Supplementary Table S1).

2.2. Molecular Detection, Isolation and Identification

Using real-time PCR, we detected pathogenic *Leptospiral* DNA in 102 out of 470 biological samples (21.70%). In total, 133 out of 347 samples (38.33%) were cultured successfully. The MAT carried out using the isolated strains as antigens against reference anti-sera showed that they were antigenically related to the Pomona serogroup. In total, 12 randomly chosen isolates (ID321, 331, 335, 340, 349, 354, 362, 385, 391, 430, 393 and 411) were subjected to serovar identification with mAbs. The reference strains Pomona and Mozdok 5621 reacted to mAb panels as expected (Table 5). Our isolates 321, 331, 335, 340, 349, 354, 362, 385, 391 and 430 reacted only to mAbs F43 C9 and F48 C6, suggesting that they belonged to the serovar Pomona of the *L. interrogans* species. The reaction profiles of isolates 393 and 411 were very similar to each other, but different from any other strain of the serogroup Pomona. If we consider an acceptable level of variability for the two dilutions used for the reaction titers, then all the mAbs, except for F48 C6 against which both isolates did not show any reactivity, indicated that they were related to the reference strain Mozdok 5621. The Tsaratsovo reference strain B 81/7 had reaction patterns for mAbs F48 C6 and F61 C7 that were similar to those of our isolates. However, the Tsaratsovo reaction patterns against mAbs F43 C9, F46 C9 and F58 C1 are unknown; therefore, they could not be compared with those of our isolates.

Table 5. Agglutination titers of Monoclonal Antibodies (mAbs) against the Pomona serogroup reference strains and the isolated samples from pigs (titers in reciprocal).

Species	Serogroup	Serovar	Strain	ID IZSLER *	mAbs				
					F43 C9	F46 C9	F48 C6	F58 C1	F61 C7
L. interrogans	Pomona	Pomona	Pomona	222	1280	0	5120	0	0
L. kirschneri	Pomona	Mozdok	5621	311	2560	10,240	2560	20,480	10,240
L. interrogans	Pomona	Pomona		321	320	0	640	0	0
L. interrogans	Pomona	Pomona		331	640	0	1280	0	0
L. interrogans	Pomona	Pomona		335	640	0	10,240	0	0
L. interrogans	Pomona	Pomona		340	320	0	2560	0	0
L. interrogans	Pomona	Pomona		349	160	0	1280	0	0
L. interrogans	Pomona	Pomona		354	640	0	5120	0	0
L. interrogans	Pomona	Pomona		362	2560	0	5120	0	0
L. interrogans	Pomona	Pomona		385	1280	0	5120	0	0
L. interrogans	Pomona	Pomona		391	2560	0	2560	0	0
L. interrogans	Pomona	Pomona		430	320	0	1280	0	0
unknown	Pomona	unclassified		393	1280	20,480	0	40,960	20,480
unknown	Pomona	unclassified		411	5120	40,960	0	40,960	20,480

*: The asterisk indicates the identification numbers of the isolates present in Istituto Zooprofilattico Sperimentale della Lombardia e dell'Emilia Romagna (IZSLER) collection.

2.3. Genotyping Analyses

2.3.1. MLST and MLVA Analyses

The genotypes of 43 randomly chosen isolates (including the 12 isolates tested for serovar determination) and the 8 biological samples (Supplementary Table S2) were determined. Using the MLST technique, 41 out of 43 isolates and all 8 DNAs extracted from biological samples were identified as ST140 and clustered with reference strain *L. interrogans*, Pomona st. Pomona (international reference strain) and st. Mezzano I (national reference strain). In total, 2 (ID 393 and 411) out of 43 isolates showed two new sequence types (STs), similar to ST117, typical of *L. kirschneri* Mozdok and also previously determined for reference strain 5621 present in the Istituto Zooprofilattico Sperimentale della Lombardia e dell'Emilia Romagna (IZSLER) collection. Compared with ST117 (pattern 13-25-15-22-33-18-23 for the loci glmU, pntA, sucA, tpiA, pfkB, mreA and caiB, respectively), the glmU gene of sample 393 showed the substitution 120 T > C, and the pntA gene had the substitution 410 A > G. The isolate 411, compared with ST117, showed the mutated pntA gene already found in isolate 393, while the glmU gene was the same as that of ST117 (allele 13). Both the sequences were submitted to the curators of the *Leptospira* database and alleles 75 and 85 were assigned to glmU and pntA, respectively. These new alleles defined two new STs, ST288 and ST289, for isolates 393 and 411, respectively.

As shown in Figure 3, 41 out of 43 isolates and all the DNA samples clustered together with the *L. interrogans* Pomona reference strains (Mezzano I and Pomona) having ST140, while the samples 393 and 411 clustered with *L. kirschneri* species near the reference strain 5621.

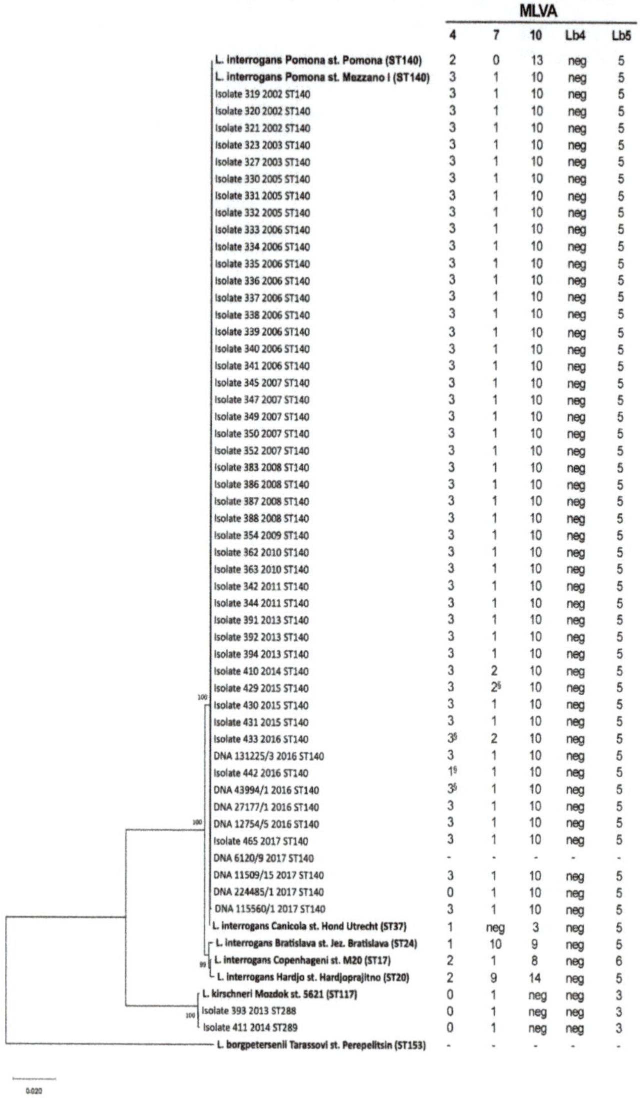

Figure 3. Phylogenetic tree based on concatenated sequences of the seven gene of the multi-locus sequence typing scheme (3111 bp) [22]. Samples are indicated with their unique IDs, the isolation year and their sequence types (STs). The names of reference strains (in bold) include the *Leptospira* species serovar and strain. The percentages of replicate trees in which the associated taxa clustered together in the bootstrap test (1000 replicates) are shown next to the branches. In the multiple-Loci Variable Tandem Repeat Analysis (MLVA) profile (on the right), the symbol "§" indicates the sequenced alleles.

Moreover, the MLVA analysis revealed that 37 out of 41 isolates, belonging to ST140, and 6 out of 8 DNA samples, had the MLVA profile 3-1-10-neg-5 for the VNTR loci 4, 7, 10, Lb4 and Lb5, respectively. This profile was identical to that of the Italian reference strain (Mezzano I), but different from the international reference strain (Pomona) profile (2-0-13-neg-5) at the loci 4, 7 and 10 (Figure 3). Three isolates belonging to ST140 (one each from 2014, 2015 and 2016) showed a new genotype (3-2-10-neg-5), which differed from strain Mezzano I and from the other samples at the VNTR7 locus. One ST140 isolate from 2017 (ID 442) harbored the new genotype 1-1-10-neg-5. One extract from 2017 (224485/1) and genotyped as ST140 showed a new VNTR profile that was 0-1-10-neg-5. The new alleles were confirmed by sequencing (Figure 3). No results were obtained for sample 6120/9, although many amplification attempts have been made for each of the five VNTR loci (Figure 3). The isolates 393 and 411 showed the same genotype found in strain Mozdok 5621, which is 0-1-neg-neg-3 for the loci 4, 7, 10, Lb4 and Lb5 respectively.

Sequencing the 16S rRNA gene confirmed that strains 393 and 411 belonged to *L. kirschneri* species, in agreement with the previous results.

2.3.2. Whole Genome Sequencing (WGS) and Core Genome MLST (cgMLST) Analysis

To clarify the genetic features of strains 393 and 411 belonging to the new STs, we performed WGS and a cgMLST analysis. Sequences of strain 393, 411, IZSLER 349/2007 and Mozdok 5621 were submitted to EBI database under the project number PRJEB36553 and are available under genome accession numbers from ERR4056312 to ERR4056315. The cgMLST analysis was performed on contigs of the strains 393, 411, IZSLER 349/2007 and Mozdok 5621 using the Bacterial Isolate Genome Sequence Database (BIGSdb) [25] (Table 6). The strains 393, 411 and IZSLER 349/2007 showed similarities with more than one cgST profile.

Table 6. Core genome (cg) MLST analysis using the Bacterial Isolate Genome Sequence Database.

	Sample Strains			
	393	**411**	**Mozdok 5621**	**IZSLER 349/2007**
Closest cgST(s) present in BIGSdb	cgST200 cgST201	cgST106 cgST111 cgST200 cgST201	cgST106	cgST373 cgST376
Loci matched with the closest cgST(s)	524/545 (96.1%)	523/545 (96%)	523/545 (96%)	539/545 (98.9%)
Clonal group *	73	73	73	5

* Based on a cut-off value of 40 allelic differences to define clonal groups, as reported by Guglielmini et al. [24]

The closest cgSTs were queried against the BIGSdb to retrieve deposited isolates with the same genotypes. The results are reported in Table 7.

Table 7. Metadata of isolates having the most similar allelic profiles to those of our samples retrieved from BIGSdb.

cgST	ST *	ID	Isolate	Species	Serogroup	Country	Serovar	Host
106	117	134	Vehlefans 2	*L. kirschneri*	Pomona	Unknown	Mozdok	Cow
111	117	140	Vehlefans 3	*L. kirschneri*	Pomona	Portugal	Mozdok	Mouse
200	117	251	61H	*L. kirschneri*	Pomona	Brazil	Mozdok	Human
201	117	252	M36/05	*L. kirschneri*	Pomona	Brazil	Mozdok	Rat
373/376	140	455	201700301	*L. interrogans*	Pomona	Italy	unknown	Cow
373/376	140	456	201700306	*L. interrogans*	Pomona	Italy	unknown	Other mammal

* Using MLST scheme 1 developed by Boonsilp [22].

The identities of strains Mozdok 5621 and IZSLER 349/2007, included in the cgST analysis as controls, were confirmed (Tables 6 and 7). The isolates 393 and 411 showed the greatest similarity (96%) with deposited strains classified as *L. kirschneri* Pomona Mozdok (Tables 6 and 7). We performed a cgMLST comparison (BIGSdb Genome Comparator [25]) among genomes of sequenced isolates (strain 5621, 393 and 411) and cgST-related strains present in the database (134|Vehlefans_2, 140|Vehlefans_3, 251|61H and 252|M36/05) (Supplementary File S1).

In total, 17 and 19 out of 545 alleles were revealed as "new" in isolates 393 and 411, respectively. In total, 14 occurred in both isolates 393 and 411 (Supplementary Table S3). The distance matrix (Figure 4a) was used to build a distance tree using SplitsTree software ver. 4.15.1 [26] (Figure 4b).

Strain	IZSLER|393	IZSLER|411	Mozdok|5621	134|Vehlefans_2	140|Vehlefans_3	251|61H	252|M36/05
IZSLER|393							
IZSLER|411	8						
Mozdok|5621	32	33					
134|Vehlefans_2	30	30	29				
140|Vehlefans_3	29	29	29	3			
251|61H	30	31	31	10	9		
252|M36/05	28	29	30	8	7	8	

(**a**)

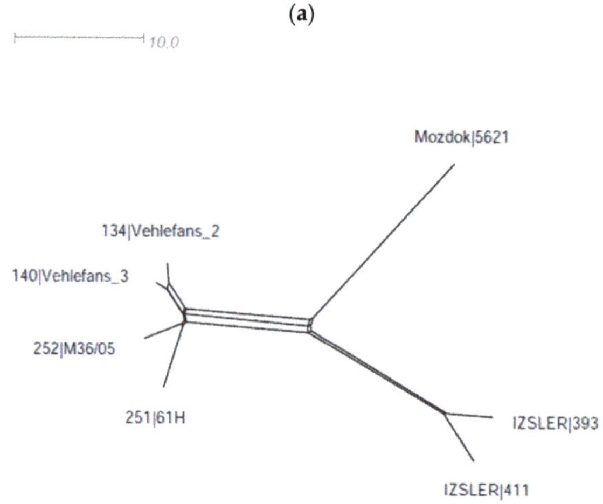

(**b**)

Figure 4. Distance matrix of the cgMLST of IZSLER isolates and cgST-related strains in the BIGSdb (**a**) and SplitsTree networks constructed using the distance values (**b**). (**a**) The color level of the distance matrix is proportional to the similarity value between strains, from green (high similarity) to red (low similarity). Incomplete loci were ignored in pairwise comparisons unless the locus was missing in an isolate. (**b**) Distance matrix visualized as SplitsTree networks.

The strains 393 and 411 were very similar, having only eight allelic differences. Compared with Mozdok 5621, they had 32 and 33 allelic mismatches, respectively (Supplementary File S1 and Figure 4) and showed from 28 to 31 differences when compared with the other serovar Mozdok strains retrieved from the database (134|Vehlefans_2, 140|Vehlefans_3, 251|61H and 252|M36/05) (Supplementary File S1 and Figure 4a). The SplitsTree network revealed three different strain clusters: the first contained the Mozdok strains retrieved from the BIGSdb (3–10 differences between each other), the second contained the IZSLER strains 393 and 411 (eight differences between each other) and the last contained Mozdok 5621, which was located on a separated branch (Figure 4b).

3. Discussion

The main objective of this study was to evaluate the epidemiology of infective *Leptospira* serovars in fattening pigs from farms in Northern Italy (Lombardy and Emilia Romagna regions). This territory has a very large pig population, with a mean of 5,300,000 animals, and contained 12,900 farms from 2007 to 2017 (data for 2002–2006 are not available), which produce pork for the entire country. In this area, the control of swine pathogens responsible for zoonoses, like *Leptospira*, is crucial to limit human and domestic animal infections, as well as to reduce the related economic losses linked to such diseases on the farms. Furthermore, the identification of the strains responsible for leptospirosis in pigs and their specific genotypes are essential for epidemiological analyses needed to adopt proper preventive measures and to better understand the etiology of the disease.

In this study, MAT, considered the gold standard and the most commonly used test, was used for serological identification. We tested 131,660 pig sera collected for 16 years. In 2002, there were 18,096 collected samples but the number progressively decreased each year to 2505 samples in 2017, except for a peak of 9330 samples in 2012. This trend may result from the strictness of the Italian national rules [9,10], in which the infection associated with a disease outbreak is only considered eradicated when all the affected animals become seronegative. This could require months or even years, and during this period, the herd is subjected to restrictive measures that are very penalizing and economically disadvantageous to the farm. Moreover, a better sampling strategy may have contributed to the reduced numbers of samples over the last few years. In fact, in the past, sampling had been carried out in all the livestock sectors, but in the last few years, perhaps for economic reasons, the trend has been to perform targeted sampling, mainly among reproductive or symptomatic animals.

In the MAT analysis, to discriminate positive and negative sera, we considered a cut-off titer of 1:100, following the guidelines indicated by the OIE [11] and commonly used for international trade. More than 80% of the positive sera (relative to 10.49% of the total samples) tested positive for one serogroup, presumably indicating the real infecting serovar and the chronic stage of infection [7]. Instead, 19.64% of the positive sera reacted simultaneously with two or more serovars, indicating both cross-reactions of serovars of the same serogroup and the acute phase of infection [4,7]. In fact, paradoxical immune response, in which the predominant serogroup in MAT is unrelated to the infecting one, occurs in the acute phase of infection meaning that 2.56% (3375 out of 133,660 sera) of all pigs examined in this study were very likely in an acute stage of infection [7].

Considering both single positive sera and multiple positive sera, our results indicated a seropositivity of 13.05%, in agreement with other previous studies [27–29]. In particular, Bertelloni found a seroprevalence of 16.6% among slaughtered swine in North–Central Italy [27], Cerri reported a prevalence of 8.85% in swine sera collected in Italy from 1995 to 2001 [28], and Chiari and colleagues found an overall prevalence of 15.28% among wild boars in Northern Italy in a five-year period (2008–2013) [29]. In Brazil, the prevalence of seropositive pigs is 16.1% [30] and in Japanese herds it is 14.4% [31]. However, a very low prevalence (1.02%) has been found in the pig population of Poland [32].

Our data revealed that a mean of 53.6% of pig farms sampled in Northern Italy tested positive for leptospirosis, in agreement with the data reported by Bertelloni and colleagues [27], who observed a leptospirosis prevalence of 52.5% in farms located in North–Central Italy from September to December 2015. Interestingly, positivity among farms showed a wide variability, from a minimum of 22.8% in 2006 to a maximum of 77.5% in 2012, in accordance with the tendency over the last few years to perform targeted sampling on suspected cases.

The most common serovars associated with swine infections in Italy are Pomona (serogroup Pomona), Tarassovi (serogroup Tarassovi), Bratislava (serogroup Australis) and Muenchen (serogroup Australis) [8]. Our results indicated that Australis is still endemic in pigs in Northern Italy, where swine acts as reservoir host for this serogroup, as in many countries and regions worldwide [4]. More than 90% of Australis-infected animals had low antibody titers (≤1:200); this is in agreement with previous studies that described serogroup Australis, specifically serovar Bratislava, as pathogenic for swine, although it usually causes a weak immune response and most of the chronically infected pigs have titers less than 1:100 [33–35]. Infections caused by Bratislava are characterized by mild clinical signs and are frequently associated with subfertility and litter-size reduction [36,37]. Unfortunately, since any Bratislava strains were successfully isolated, there is no data that allow us to define the specific infective serovar present on Australis-positive farms. It is probable that these infections were largely caused by *Leptospira* Bratislava, which is widespread among hedgehogs (*Erinaceus europaeus*) [38] and was found in wild boars in Parma Apennines in Italy [39]. However, we cannot exclude a priori that there are other causal agents, such as the Muenchen and Lora serovars. In fact, these serovars, together with Bratislava, have been isolated from the genital tracts and kidneys of sows that have abortions, from aborted fetuses and from boars in the USA [40], the Netherlands [41] and Northern Ireland [42,43]. Serogroup Australis has emerged as the major swine-maintained cause of *Leptospira* infections, but its epidemiological role remains poorly understood owing to isolation difficulties [44]. The isolation of serovar Bratislava still represents a challenge in the diagnosis of leptospirosis and, in fact, in contrast to the high seroprevalence reported worldwide [3,45–47], *L. interrogans* Bratislava has only been recovered from swine in a few countries, such as the United Kingdom [48], Germany [44], the USA [40,49,50], Vietnam [51] and Brazil [36]. Even in this study, we did not isolate any Bratislava strains. This may result from the quality and composition of the medium used, as well as the low bacterial load in kidneys, the organ used for the culture isolation [52]. In fact, Ellis et al. [52] found a higher isolation frequency for the Australis serogroup (Bratislava and Muenchen) from genital organs compared with kidneys. Therefore, improvements in the quality and composition of the isolation medium, as well as sampling procedures, must be considered.

Pomona, the second most predominant serogroup in this work, is among the most common serogroup isolated from swine worldwide [53–55], and many strains of this serogroup are adapted to swine, which is recognized as a maintenance host [3]. Pigs infected with serogroup Pomona often experience abortions, stillbirths or the births of weak or ill piglets, and the infections result in subsequent decreased reproductive performances. Adult non-pregnant animals are usually asymptomatic carriers. Unfortunately, no data regarding the clinical symptoms of the analyzed pigs in this study were available. Swine infections caused by Pomona strains have been observed in Europe, particularly in its eastern and southern regions [56,57]. Recently, Bertelloni and colleagues [27] highlighted a recurrence of serogroup Pomona in pigs in North–Central Italy, revealing a Pomona serogroup seropositivity of 18% in the screened farms and of 8% in analyzed sera. Apart from swine, Pomona also causes clinically and economically significant infections in cattle, sheep and horses in many countries, resulting particularly in reproductive problems and equine uveitis [58]. A study from New Zealand reported that 31% of people working or residing on or within a 50-km radius of pig farms had microscopic agglutination titers of 1:24 or greater to one or more serovars of *L. interrogans*, with the majority being to Pomona [59]. Moreover, the ST140 genotype, the same as in our infected pigs, has been associated with symptomatic human leptospirosis in New Caledonia, Australia and Sri Lanka (this publication made use of the *Leptospira* MLST website https://pubmlst.org/Leptospira/ sited at the University of Oxford [25]. The development of this site has been funded by the Wellcome Trust).

For Tarassovi, historical data supported the widespread presence of this serogroup among the Italian pig population. In 1990, Tagliabue reported the isolation of four Tarassovi strains from swine kidneys [60], while in 1993, Scanziani and colleagues [37] described mild renal lesions in pigs affected by Tarassovi. Additionally, in 1995, Tagliabue and Farina reported a seroprevalence of 6.8% among the pig population, using a cut-off titer of 1:400 [38]. The identification of Tarassovi as the infective serogroup of swine in the past was corroborated by its presence in the vaccine formulation, indicating the need to protect hosts from this infective agent. In accordance with the observations already made by Tagliabue and Farina in 1995 [38], which noted a progressive decline in the Tarassovi frequency during the last few years, our study found that, in Italy, the infections caused by Tarassovi are decreasing, being responsible for the positivity of only 0.16% of the total samples tested. The infections caused by Tarassovi in swine have become incidental and probably result from contact with wild animals, including turtles, which act as a maintenance host for this serovar [61].

Among the minor serogroups found in tested pigs, we payed particular attention to some of them, discussed below. First of all, serogroup Icterohaemorrhagiae whose low prevalence could indicate the use of good rodent control measures inside and around the tested pig farms [62]. Moreover, the low prevalence observed for Grippotyphosa, Sejroe and Canicola were in agreement with historical data that described these serogroups as causes of incidental infection transmitted to pigs from wildlife hosts (serogroup Grippothyphosa [4]), from dogs (serogroup Sejroe serovar Sejroe [63] and serogroup Canicola serovar Canicola [64]), from bovines (serogroup Sejroe serovar Hardjo [4]) or small rodents (serogroup Sejroe serovar Saxkoebing [65]).

Through genotyping techniques we found that all the isolated and extracted samples belonged to the Pomona serogroup but, interestingly, from 2014, new Pomona genotypes started to circulate in Northern Italy. Unfortunately, the Sequence Type provided by MLST analysis (ST140) being associated with isolates characterized as belonging to serogroups Pomona serovar Pomona, serogroup Pyrogenes serovar Guaratuba and serogroup Grippothyphosa (data from BIGSdb) did not permit any discrimination at the serogroup and at the serovar level, this ST In this case, the information provided by MLST genotyping is limited to species discrimination (i.e., *L. interrogans*). However, we should distinguish two different situations: in the case of isolated strains, since serological analysis by polyclonal anti-sera and by mAbs were possible, we were able to distinguish their serovar (Pomona) and serogroup (Pomona). In the case of DNA samples, we cannot assume that they are definitively serovar Pomona based only on the MLST results. However, we can assume that six of the eight samples belonged to serogroup Pomona serovar Pomona because their MLVA profiles were identical to that of strain Mezzano I.

Interestingly, two isolates showed new mAb responses and new STs. Their reactions to the mAbs specific for the Pomona serogroup were very similar to each other but different from any of the known serogroup Pomona strain profiles. Their pattern was compatible with serovar Tsaratsovo, which was first isolated from harvest mice (*Micromys minutus*) in 1962, in the Plovdiv District of Bulgaria by Ivanov and later found in rodents in some parts of Europe [66]. Nevertheless, only mAbs F48 C6 and F61 C7, included in our panel, were tested on this serovar. Therefore, these data are insufficient to determine their serovars states. However, from an epidemiological point of view, this serovar has never been found in Italy; thus, it is very improbable that Tsaratsovo exists among the Italian pig population. Genotyping analyses of these strains indicated that they had the same VNTR pattern of Mozdok st. 5621 and belonged to ST288 and ST289, which had not been described previously, and are similar to ST117, which was found in *L. kirschneri* Mozdok st. 5621. In addition, the cgMLST analysis revealed that the most similar strains are classified as *Leptospira*, species *kirschneri*, serovar Mozdok. Definitely, MLST and cgMLST analyses excluded their belonging to the Tsaratsovo serovar, because strain Tsaratsovo B 81/7 was characterized as ST115 and cgST 245. We can also exclude Altodouro and Kunming, the other serovars included in *L. kirschneri* species and the Pomona serogroup, because they showed ST100 and ST70, respectively, with none of the seven loci being in common with those of ST288 and ST289. Therefore, we can assert that they are very similar to *L. kirschneri* Mozdok, but accurate descriptions of their genetic contents in comparison with serovar Mozdok are still being compiled. From an epidemiologic point of view, several studies have identified serovar Mozdok across Europe [67–71] in dogs [72] and in wild rodents [70,73,74]. It has been described also as a possible cause of abortions in cattle [75] and in pigs [69,73,74] and has been responsible of human leptospirosis in Cuba [76].

Owing to the advent of next-generation sequencing and the publication of the cgMLST scheme, the resolution powers of sequence-based analyses on *Leptospira* are increasing but the definitions of allelic distance cut-offs for the differentiation of strains are still open and in urgent need of cgMLST classification.

Our study, which provides information on the *Leptospira* serovars present in the Italian pig population, is also useful for implementing a successful diagnostic and vaccination program. Commercial *Leptospira* vaccines are available globally for cattle, pigs and dogs, but vaccination has proven to be only partially effective, owing in part to the serovar-restricted nature of vaccine-induced immunity and the potential presence of local serovars other than those included in the vaccines. In Italy, prior to 2010, a trivalent vaccine against Bratislava (serogroup Australis), Pomona (Pomona) and Hyos (Tarassovi) was available for swine. It was used in breeding boars, in non-pregnant sows and in pregnant sows before the octave week of gestation. An unpublished survey carried out by the National Centre of Leptospirosis of IZSLER revealed that among 333 pig farms in Northern Italy during the period February 2004–December 2005, only 19% of the sampled farms adopted the vaccination practice. The reasons for this limited use of vaccination against leptospirosis in pigs included the risk that this disease is poorly understood by farmers and that leptospirosis is a secondary problem in the farm life/economy compared with other more virulent and economically important diseases. Nevertheless, since MAT does not discriminate between vaccination titers and titers due to exposure and because the vaccination status of pigs tested in this study prior to 2010 were unknown, the risk of having an altered number of positive pigs due to the interference of vaccination should be considered. About this aspect, the previously mentioned survey reported that in the farms where vaccination against *Leptospira* was practiced, 41% of pigs resulted as seropositive, while where it was not practiced seropositive animals were equal to only 9.8%. However, when considering this situation, it highlighted various scenarios: in the case of Australis, the interference of vaccination immunity was significant because the percentage of seropositives increased from 6.1% (of unvaccinated farms) to 33.9% (of vaccinated farms), while for Pomona and Tarassovi, the interference level was lower, going from 2.4% to 5.4% and from 1.2% to 5.4% for unvaccinated and vaccinated farms, respectively (data not published). Based on these data, we can exclude the interference of vaccination on our seropositivity results, given that the overall seropositivity and the seropositivity levels related to Australis, Pomona and Tarassovi (13.1%, 8.1%, 1.9% and 0.2%, respectively) were in agreement with the data from the unvaccinated farms reported in the survey (9%, 6.1%, 2.4% and 1.2%). Furthermore, in this study the trends in the seropositivity of Australis, Pomona and Tarassovi did not reveal any remarkable changes after the end of vaccine use in Italy occurred in 2010, which provided further evidence for our statement.

Considering the serological data obtained and the frequency, in particular, of the Australis and Pomona serogroups, the use of vaccination in pig herds is highly advisable. In this regard, in March 2019, a vaccine has been registered in Italy. This vaccine contains nine serovars (Bananal, Bratislava, Canicola, Copenhageni, Gryppotyphosa, Icterohaemorrhagiae Pomona, Tarassovi, Vughia) and is active against six serogroups (Australis, Canicola, Gryppotyphosa, Icterohaemorrhagiae, Pomona, Tarassovi). As reported by Jacobs et al. [77], this vaccine can be safely used in gilts and sows and induces significant protection for the duration of at least one year. Furthermore, after a challenge with serovar Pomona, it induces protection against clinical signs, leptospiraemia and foetal death [78]. Its use can be considered a valid aid in limiting the infection among pig herds in our territory.

4. Materials and Methods

4.1. Sampling

A total of 133,660 sera, 347 kidneys and 470 biological samples (from kidneys and unspecified tissues, as well as urine) were included in this study. These samples were collected over 16 years (from 2002 to 2017) from 4715 pig farms located in Northern Italy (Lombardy and Emilia Romagna regions) during the routine diagnostic activity. In these farms, testing for *Leptospira* infections was previously included in the disease-monitoring programs. The samples were collected and analyzed by the National Reference Centre for Animal Leptospirosis at the Istituto Zooprofilattico Sperimentale della Lombardia ed Emilia Romagna (IZSLER) located in Brescia (Italy). In this study, data already published by Tagliabue [61] collected in 2010 and 2011, were also included. In Table 8 we report the numbers of collected serum samples and farms.

Table 8. Numbers of serum samples analyzed by MAT and farms involved in this study.

Year	Number of Serum Samples	Number of Farms
2002	18,096	452
2003	15,161	454
2004	10,881	528
2005	11,524	478
2006	11,296	613
2007	9474	332
2008	8059	265
2009	7693	246
2010	5841	225
2011	4825	202
2012	9330	298
2013	4029	135
2014	4306	130
2015	4055	128
2016	4585	128
2017	2505	101
Total	131,660	4715

Starting from 2008, 347 well-preserved samples were submitted for culture-based causal agent isolation. Since 2009, when molecular methods for the routine diagnosis of leptospirosis were introduced, 470 clinical samples were analyzed using a LipL32-based real-time PCR assay.

To cover the entire analysis period, 43 randomly isolated samples and DNA extracted from 8 biological samples were submitted to genotyping analysis (Supplementary Table S2). For the isolation and genotyping procedures, we selected only one positive sample per farm, assuming that all the positive animals of the same herd were infected with the same strain.

4.2. Serological Tests of Serum Samples

Sera were examined for the presence of antibodies against pathogenic *Leptospira* using the MAT in accordance with OIE standards [11]. For MAT testing, live cultures of eight reference strains of *Leptospira* were used (Table 9). They were cultivated at 30 ± 1 °C in EMJH (Ellinghausen, McCullough, Johnson, Harris) medium enriched with bovine serum albumin supplement (10% v/v) [79,80]. Sera were pretested at the final dilution of 1/100. Sera with 50% agglutination were retested to determine an endpoint using dilutions of sera beginning at 1/100 through to 1/6400. Serum samples with the widely accepted minimum significant titer of 100 (reciprocal of the final dilution of serum with 50% agglutination) were assessed positive.

Table 9. Panel of eight *Leptospira spp.* used as live antigens for MAT.

Species	Serogroup	Serovar	Strain
L. interrogans	Australis	Bratislava	Riccio 2
L. borgpetersenii	Ballum	Ballum	Mus 127
L. interrogans	Canicola	Canicola	Alarik
L. kirschneri	Grippotyphosa	Grippotyphosa	Moskva V
L. interrogans	Icterohaemorrhagiae	Copenhageni	Wijnberg
L. interrogans	Pomona	Pomona	Pomona
L. interrogans	Sejroe	Hardjo	Hadjoprajitno
L. borgpetersenii	Tarassovi	Tarassovi	Mitis-Johnson

4.3. DNA Extraction and Real-Time PCR Detection

DNA was extracted from 0.5–4.0 mL of urine or tissue homogenate using the PureLink Genomic DNA kit (Invitrogen, Paisley, UK) according to the manufacturer's instructions. An internal control DNA (0.1 μL per μL of elution volume) was added to the digestion buffer. A Taqman-based PCR assay targeting the lipL32 gene was used to detect pathogenic leptospires using primers described previously [17]. The PCR was performed in a 25-μL final volume, using 5 μL of extracted DNA, 5 μL of 5× Mastermix Quantifast (Quantifast Pathogen + IC Kit, Qiagen, Hilden, Germany), 700 nM of primers and 200 nM of the probe. All extraction session included a negative control (water) and all amplification sessions included both a negative (water) and a positive control (DNA of *Leptospira interrogans* Pomona). The assay was performed on a Bio-Rad CFX96 System using the following thermal conditions: a holding stage of 95 °C for 5 min, and 45 cycles of 95 °C for 15 s and 60 °C for 30 s. Supplementary Table S4 reported the interpretation criteria of Real-time PCR results.

4.4. Isolation of Leptospira from Swine Kidneys

For isolation, 1 g of tissue was diluted, in a plastic bag, in 9 mL of EMJH added with 5-fluoruracil (selective EMJH) [79,80] and homogenized. This suspension was then filtered and collected in a sterile tube and diluted in a four ten-fold serial dilutions (10^{-2} to 10^{-5}) performed in semisolid selective EMJH (added with 5-fluoruracil and agar). The cultures were incubated at 30 ± 1 °C and observed under dark-field microscopy weekly for up to 6 months. In case of contamination, the cultures were filtrated through a 0.22 μm sterile syringe filter and sub-cultured in fresh semisolid selective EMJH medium.

4.5. Characterization of Pig Isolates Using MAT

The serogroups of the isolates were determined by MAT using a panel of eight polyclonal anti-sera against the eight serovars described in Table 9. A high rate of agglutination with a particular antiserum was used to identify the presumptive serogroup of the strain [81].

Serovar classification was achieved by MAT using panels of mAbs directed against polysaccharide antigens, specific to the serogroup identified. A battery of five mAbs (F43 C9, F46 C9, F48 C6, F58 C1 and F61 C7), which react with serovars of the serogroup Pomona [82,83], were used to determine the serovar states of our isolates. These mAbs and the polyclonal anti-sera were previously purchased from the OIE Leptospirosis Reference Centre, Royal Tropical Institute (KIT), (Amsterdam, The Netherlands), and the protocol was in accordance to the standard serological methods used in this reference laboratory. The reference strains *L. interrogans* Pomona st. Pomona and *L. kirschneri* Mozdok st. 5621, provided by KIT, were used as controls for serovar identification. To interpret the results, we referred to the expected maximum dilution titer provided by the Leptospirosis Reference Centre of KIT, as reported in Table 10.

Table 10. Maximum dilution titer patterns of the serovars in the Pomona serogroup against the available mAbs provided by the Leptospirosis Reference Centre, Royal Tropical Institute, Amsterdam, The Netherlands.

Species	Serogroup	Serovar	Strain	mAbs				
				F43 C9	F46 C9	F48 C6	F58 C1	F61 C7
L. interrogans	Pomona	Pomona	Pomona	1,280	0	5120	0	0
L. kirschneri	Pomona	Mozdok	5621	2,560	10,240	2560	20,480	10,240
L. kirschneri	Pomona	Altodouro	RIM 139	80	20,480	5120	20,480	20,480
L. kirschneri	Pomona	Tsaratsovo	B 81/7	nt	nt	0	nt	20,480
L. kirschneri	Pomona	Kunming	K5	640	2560	2560	640	2560
L. borgpetersenii	Pomona	Kunming	200901118	nt	nt	nt	nt	nt
L. interrogans	Pomona	Proechymis	1161 U	80	2560	640	40,960	10,240
L. santarosai	Pomona	Tropica	CZ 299	640	10,240	5120	1280	10,240

nt: not tested.

4.6. Genotyping

4.6.1. MLST

To genotype leptospires we used the scheme proposed by Bonsilp in 2013 [22], based on the seven housekeeping genes UDP-N-acetylglucosamine pyrophosphorylase (glmU), UDP-N-acetylglucosamine pyrophosphorylase (pntA), 2-oxoglutarate dehydrogenase E1 component (sucA), triosephosphate isomerase (tpiA), 1-phosphofructokinase (pfkB), rod shape-determining protein rodA (mreA) and acyl-CoA transferase/carnitine dehydratase (caiB). For the isolates, these loci were amplified using a KAPA2G Robust HotStart PCR kit (Kapa Biosystems Resnova, Roma, Italy) in a 25-µL total volume containing 0.4 µM each primer, 2.5 µL of boiled culture and MgCl2 at the following concentrations: 1.5 mM for mreA, pfkB, pntA, caiB and glmU; 2.5 mM for sucA; and 3.5 mM for tpiA. Temperature cycling was performed as follows: 1 cycle at 95 °C for 15 min, 35 amplification cycles of 95 °C, 55 °C for 30 s and 72 °C for 1 min, followed by a final elongation at 72 °C for 10 min. To genotype the DNA extracted from biological samples, we used the protocol described by Weiss and colleagues [23], consisting of one PCR reaction, similar to that described for isolates, but performed using 5 µL of extracted DNA, followed by a second amplification reaction, that used nested primers to improve the sensitivity. The nested PCR was performed in 25-µL reactions containing 5 pmol of each primer and 2 µL of the first-round PCR product. Cycling conditions were as follows: 95 °C for 10 min, 5 cycles of 95 °C for 30 s, 46 °C for 30 s and 72 °C for 30 s. In addition, 10 cycles, in which the annealing temperature increased by 1°C per cycle, and 20 cycles with an annealing temperature of 56 °C, were performed. A final extension step at 72 °C for 7 min was performed. The presence of PCR products was verified on 2% agarose gels stained with EuroSafe Nucleic Acid Stain (Euroclone, Milan, Italy). PCR products were purified using the NucleoSpin® Gel and PCR Clean-up (Macherey-Nagel, Düren, Germany) or with exonuclease I and FastAp Thermosensitive Phospahatase alkaline (Thermo Fisher Scientific, Waltham, MA, USA) according to the manufacturer's instructions. Cycle sequencing reactions were performed using the BigDye® Terminator Cycle Sequencing kit version 1.1 (Applied Biosystems, Foster City, CA, USA). Reactions were purified using a BigDye® XTerminatorTM Purification kit Thermo Fisher Scientific, Waltham, MA, USA) and sequenced on a Genetic Analyzer 3500xl sequencer (Thermo Fisher Scientific) according to the manufacturer's instructions. Nucleotide sequences were assembled using the SeqMan module of the Lasergene sequencing analysis software package (DNASTAR, Inc., Madison, WI, USA). Assembled sequences were trimmed and aligned to allele reference sequences downloaded from the Bacterial Isolate Genome Sequence Database (BIGSdb) [25] (https://pubmlst.org/Leptospira/), to assign allele numbers to all seven loci. For strain identification, allelic profiles were queried against the *Leptospira* BIGSdb.

The phylogenetic tree was built using the concatemer of the seven MLST genes linked in the followed order: glmU-pntA-sucA-tpiA-pfkB-mreA-caiB using MEGA7 [84]. The phylogeny was inferred using the Neighbor-Joining method and by calculating genetic distances using the p-distance method.

4.6.2. VNTR Analysis

Five discriminatory loci (VNTR-4, VNTR-7, VNTR-10, VNTR-Lb4 and VNTR-Lb5) were used to characterize the isolates and the extracted DNA samples as described by Salaün and colleagues in 2006 [21]. Briefly, 5 µL of boiled isolate or extracted DNA was added to 20 µL of the KAPA2G Robust HotStart PCR kit (Kapa Biosystems Resnova) reaction mixture that contained 10 pmol of each primer. The PCR was carried out as follows: 95 °C for 15 min and 35 cycles of 95 °C for 30 s, 55 °C for 30 s and 72 °C for 1 min. An additional extension for 10 min at 72 °C was added to the end of the run. The PCR products were analyzed on 2% agarose gels stained with EuroSafe Nucleic Acid Stain (Euroclone), and the molecular weights were estimated by comparison with a 100-bp DNA ladder (Thermo Fisher Scientific). If the length of PCR product was unclear, we sequenced it using the same sequencing protocol described for the MLST analysis.

4.6.3. Genomic DNA Extraction from Cultured Strains

Genomic DNA of two *Leptospira* isolates from pigs belonging to new Sequence Types (ID 393 and 411) were extracted from the liquid EMJH cultures using the DNeasy Blood and Tissue kit (Qiagen), according to the manufacturer's instructions. Extracted DNA was successfully used for species confirmation and for WGS. One strain isolated from pig previously identify as *L. interrogans* servar Pomona, "strain IZSLER 349/2007", and the reference strain Mozdok 5621 were included as controls for the cgMLST analysis.

4.6.4. Species Confirmation

The species assignments of the isolates 393 and 411 were confirmed by sequencing portions of their 16S rRNA genes using the Microseq 500 16S rDNA PCR kit (Applied Biosystems, Foster City, CA, USA). The obtained sequences were compared with a controlled and validated reference database (MicroseqID; Applied Biosystems) and through BLAST-based analyses to determine the species of the *Leptospira* strains.

4.6.5. WGS and the cgMLST Analysis

WGS was performed on the four genomic DNAs extracted from cultured strains (ID 393, 411, IZSLER 349/2007 and strain Mozdok 5621) using an Illumina NextSeq (Illumina, San Diego, CA, USA) platform with a paired-end 150× 2-bp run starting from a genomic library prepared with a Nextera DNA Flex kit (Illumina). Raw reads were filtered using Trimmomatic ver. 0.38 [85], assembled using SPAdes Assembler ver. 3.9.0 [86] and evaluated using QUAST ver. 4.2 [87]. The resulting contigs were taxonomically classified using Kaiju software [88] and only the *Leptospira* contigs were used for further analyses. The draft genomes of samples 393, 411, IZSLER 349/2007 and strain Mozdok 5621 were used to perform cgMLST analyses with BIGSdb [25], using the cgMLST scheme recently developed by Guglielmini et al. [24].

Ethical statement: The study was exempt of ethical approval procedures because animal samplings were performed during the routinely diagnostic procedures in naturally infected farms.

5. Conclusions

In this study, an intensive serological survey evidenced that Australis and Pomona were the most serogroups causing leptospirosis in pigs in Italy. Molecular analyses revealed that a stable genotype of strains belonging to *L. interrogans* serogroup Pomona had been circulating for 10 years (from 2002 until 2013) and interestingly, *L. interrogans* strains collected from 2014 onwards had various heterogeneous genetic profiles. Furthermore, new strains belonging to *L. kirschneri* species have been identified for the first time. Since an accurate identification of the infective strain is crucial for an adequate vaccine formulation, these findings provide an important contribution for addressing prevention and intervention strategies to reduce infection risks on farms.

Supplementary Materials: The following are available online at http://www.mdpi.com/2076-0817/9/5/332/s1, Table S1: Trends of outbreaks observed in 2002–2017 (only single positivities were considered), Table S2: Samples submitted for genotyping analysis, Table S3: Loci showing new alleles in samples 393 and 411. Table S4: Interpretation of Real-time PCR results. Common loci having the same new alleles in samples 393 and 411 are shown in bold, File S1: Comparison of allelic profiles resulting from the cgMLST analysis.

Author Contributions: Conceptualization, M.B.B., C.B. and M.D.; methodology, C.B., A.P. and E.S.; formal analysis, C.B.; investigation, C.B., M.B.B., M.D. and S.T.; resources, M.D. and M.B.B.; data curation, C.B. and M.D.; writing—original draft preparation, C.B. and M.B.B.; writing—review and editing, M.B.B., M.D., S.T. and E.S.; project administration, M.B.B. and M.D.; funding acquisition, M.B.B. and M.D. All authors have read and agreed to the published version of the manuscript.

Funding: This work was funded by the Italian Ministry of Health (project PRC2013006 and project PRC2017016).

Acknowledgments: We thank Arturo Scalvenzi, Marcello Fin and Eufrasia Peroni Ammaturo for their skilled technical assistance with the serology and bacterial isolation, and Anna Mangeli and Daniela Loda for their expert technical support for molecular and genotyping techniques. We are very grateful to Janjira Thaipadungpanit

(Mahidol-Oxford Tropical Medicine Research Unit, Faculty of Tropical Medicine, Mahidol University, Bangkok, Thailand) for assigning new allelic numbers and sequence types to our Italian strains. Thank you also to Science Docs for the English language editing of this manuscript.

Conflicts of Interest: The authors declare no conflict of interest. The funders had no role in the design of the study; in the collection, analyses, or interpretation of data; in the writing of the manuscript, or in the decision to publish the results.

References

1. Bharti, A.R.; Nally, J.E.; Ricaldi, J.N.; Matthias, M.A.; Diaz, M.M.; Lovett, M.A.; Levett, P.N.; Gilman, R.H.; Willig, M.R.; Gotuzzo, E.; et al. Leptospirosis: A zoonotic disease of global importance. *Lancet Infect. Dis.* **2003**, *3*, 757–771. [CrossRef]
2. Adler, B.; de la Pena Moctezuma, A. Leptospira and Leptospirosis. *Vet. Microbiol.* **2010**, *140*, iv–v. [CrossRef] [PubMed]
3. Ellis, W.A. Leptospirosis. In *Diseases of Swine*; Blackwell: Oxford, UK, 1999; pp. 483–554.
4. Strutzberg-Minder, K.; Tschentscher, A.; Beyerbach, M.; Homuth, M.; Kreienbrock, L. Passive surveillance of Leptospira infection in swine in Germany. *Porc. Heal. Manag.* **2018**, *4*, 10. [CrossRef] [PubMed]
5. Burnstein, T.; Baker, J.A. Leptospirosis in swine caused by Leptospira pomona. *J. Infect. Dis.* **1954**, *94*, 53–64. [CrossRef]
6. Yamaguchi, T.; Higa, N.; Okura, N.; Matsumoto, A.; Hermawan, I.; Yamashiro, T.; Suzuki, T.; Toma, C. Characterizing interactions of *Leptospira interrogans* with proximal renal tubule epithelial cells. *BMC Microbiol.* **2018**, *18*, 64. [CrossRef]
7. Levett, P.N. Leptospirosis. *Clin. Microbiol. Rev.* **2001**, *14*, 296–326. [CrossRef]
8. Nassuato, C.; Cominardi, P.; Tagliabue, S.; Pennelli, D. Gestione di un focolaio di *Leptospira interrogans* variante Pomona in un allevamento suino da ingrasso 1998. *Osservatorio* **2006**, *9*, 4–9.
9. O.M. 4 September 1985, Profilassi delle leptospirosi animali. *Gazz. Uff.* **1985**, *226*. Available online: https://gestione.izsler.it/izs_bs/allegati/5753/1_OM_04091985.pdf (accessed on 25 April 2020).
10. D.P.R. 320/1954. Regolamento di polizia veterinaria. *Gazz. Uff.* **1954**, *142*. Available online: http://www.salute.gov.it/imgs/C_17_normativa_925_allegato.pdf (accessed on 25 April 2020).
11. Leptospirosis. In *OIE Manual of Diagnostic Tests and Vaccines for Terrestrial Animals*, 8th ed.; World Organisation for Animal Health: Paris, France, 2018; Chapter 3.1.12; pp. 1–3, 503–516.
12. Wuthiekanun, V.; Chierakul, W.; Limmathurotsakul, D.; Smythe, L.D.; Symonds, M.L.; Dohnt, M.F.; Slack, A.T.; Limpaiboon, R.; Suputtamongkol, Y.; White, N.J.; et al. Optimization of culture of Leptospira from humans with leptospirosis. *J. Clin. Microbiol.* **2007**, *45*, 1363–1365. [CrossRef]
13. Smythe, L.D.; Smith, I.L.; Smith, G.A.; Dohnt, M.F.; Symonds, M.L.; Barnett, L.J.; McKay, D.B. A quantitative PCR (TaqMan) assay for pathogenic Leptospira spp. *BMC Infect. Dis.* **2002**, *2*, 13. [CrossRef] [PubMed]
14. Levett, P.N.; Morey, R.E.; Galloway, R.L.; Turner, D.E.; Steigerwalt, A.G.; Mayer, L.W. Detection of pathogenic leptospires by real-time quantitative PCR. *J. Med. Microbiol.* **2005**, *54*, 45–49. [CrossRef] [PubMed]
15. Palaniappan, R.U.M.; Chang, Y.F.; Chang, C.F.; Pan, M.J.; Yang, C.W.; Harpending, P.; McDonough, S.P.; Dubovi, E.; Divers, T.; Qu, J.; et al. Evaluation of lig-based conventional and real time PCR for the detection of pathogenic leptospires. *Mol. Cell. Probes* **2005**, *19*, 111–117. [CrossRef] [PubMed]
16. Roczek, A.; Forster, C.; Raschel, H.; Hörmansdorfer, S.; Bogner, K.H.; Hafner-Marx, A.; Lepper, H.; Dobler, G.; Büttner, M.; Sing, A. Severe course of rat bite-associated Weil's disease in a patient diagnosed with a new Leptospira-specific real-time quantitative LUX-PCR. *J. Med. Microbiol.* **2008**, *57*, 658–663. [CrossRef]
17. Stoddard, R.A.; Gee, J.E.; Wilkins, P.P.; McCaustland, K.; Hoffmaster, A.R. Detection of pathogenic Leptospira spp. through TaqMan polymerase chain reaction targeting the LipL32 gene. *Diagn. Microbiol. Infect. Dis.* **2009**, *64*, 247–255. [CrossRef]
18. Slack, A.T.; Symonds, M.L.; Dohnt, M.F.; Smythe, L.D. Identification of pathogenic Leptospira species by conventional or real-time PCR and sequencing of the DNA gyrase subunit B encoding gene. *BMC Microbiol.* **2006**, *6*, 95. [CrossRef]
19. Noubade, R.; Krishnamurthy, G.V.; Murag, S.; Venkatesha, M.D.; Krishnappa, G. Differentiation of pathogenic and saprophytic leptospires by polymerase chain reaction. *Indian J. Med. Microbiol.* **2002**, *20*, 33.

20. Kositanont, U.; Rugsasuk, S.; Leelaporn, A.; Phulsuksombati, D.; Tantitanawat, S.; Naigowit, P. Detection and differentiation between pathogenic and saprophytic Leptospira spp. by multiplex polymerase chain reaction. *Diagn. Microbiol. Infect. Dis.* **2007**, *57*, 117–122. [CrossRef]
21. Salaün, L.; Mérien, F.; Gurianova, S.; Baranton, G.; Picardeau, M. Application of Multilocus Variable-Number Tandem-Repeat Analysis for Molecular Typing of the Agent of Leptospirosis. *J. Clin. Microbiol.* **2006**, *44*, 3954–3962. [CrossRef]
22. Boonsilp, S.; Thaipadungpanit, J.; Amornchai, P.; Wuthiekanun, V.; Bailey, M.S.; Holden, M.T.G.; Zhang, C.; Jiang, X.; Koizumi, N.; Taylor, K.; et al. A single multilocus sequence typing (MLST) scheme for seven pathogenic Leptospira species. *PLoS Negl. Trop. Dis.* **2013**, *7*, e1954. [CrossRef]
23. Weiss, S.; Menezes, A.; Woods, K.; Chanthongthip, A.; Dittrich, S.; Opoku-Boateng, A.; Simuli, M.; Chalke, V. An Extended Multilocus Sequence Typing (MLST) Scheme for Rapid Direct Typing of Leptospira from Clinical Samples. *PLoS Negl. Trop. Dis.* **2016**, *10*, e0004996. [CrossRef] [PubMed]
24. Guglielmini, J.; Bourhy, P.; Schiettekatte, O.; Zinini, F.; Brisse, S.; Picardeau, M. Genus-wide Leptospira core genome multilocus sequence typing for strain taxonomy and global surveillance. *PLoS Negl. Trop. Dis.* **2019**, *13*, e0007374. [CrossRef]
25. Jolley, K.A.; Maiden, M.C.J. BIGSdb: Scalable analysis of bacterial genome variation at the population level. *BMC Bioinform.* **2010**, *11*, 595. [CrossRef] [PubMed]
26. Huson, D.H.; Bryant, D. Application of Phylogenetic Networks in Evolutionary Studies. *Mol. Biol. Evol.* **2006**, *23*, 254–267. [CrossRef] [PubMed]
27. Bertelloni, F.; Turchi, B.; Vattiata, E.; Viola, P.; Pardini, S.; Cerri, D.; Fratini, F. Serological survey on Leptospira infection in slaughtered swine in North-Central Italy. *Epidemiol. Infect.* **2018**, *146*, 1275–1280. [CrossRef] [PubMed]
28. Cerri, D.; Ebani, V.V.; Fratini, F.; Pinzauti, P.; Andreani, E. Epidemiology of leptospirosis: Observations on serological data obtained by a "diagnostic laboratory for leptospirosis" from 1995 to 2001. *New Microbiol.* **2003**, *26*, 383–389.
29. Chiari, M.; Figarolli, B.M.; Tagliabue, S.; Alborali, G.L.; Bertoletti, M.; Papetti, A.; D'Incau, M.; Zanoni, M.; Boniotti, M.B. Seroprevalence and risk factors of leptospirosis in wild boars (*Sus scrofa*) in northern Italy. *Hystrix* **2016**, *27*.
30. Valença, R.M.B.; Mota, R.A.; Castro, V.; Anderlini, G.A.; Júnior, J.W.P.; Brandespim, D.F.; Valença, S.R.F.A.; Guerra, M.M.P. Prevalence and Risk Factors Associated with *Leptospira* spp. Infection in Technified Swine Farms in the State of Alagoas, Brazil Risk Factors Associated with *Leptospira* spp. in Swine Farms. *Transbound. Emerg. Dis.* **2013**, *60*, 79–86. [CrossRef]
31. Naito, M.; Sakoda, Y.; Kamikawa, T.; Nitta, Y.; Hirose, K.; Sakashita, M.; Kurokawa, S.; Kida, H. Serological evidence of leptospiral infection in pig populations in different districts in Japan. *Microbiol. Immunol.* **2007**, *51*, 593–599. [CrossRef]
32. Wasinski, B. Occurrence of Leptospira serovars in pigsin the years 2002–2003. *Med. Weter.* **2005**, *61*, 46–49.
33. Higgins, R.; Champagne, M.J. La leptospirose porcine. *Le Médicin Véterinaire du Québec* **1993**, *23*, 7–12.
34. Hanson, L.E. Bratislava in swine. *In Proc Am Assoc Swine Pr.* **1987**, 85–91.
35. Miller, D.A.; Wilson, M.A.; Owen, W.J.; Beran, G.W. Porcine Leptospirosis in Iowa. *J. Vet. Diagnostic Investig.* **1990**, *2*, 171–175. [CrossRef] [PubMed]
36. Hamond, C.; Martins, G.; Loureiro, A.P.; Bremont, S.; Medeiros, M.A.; Bourhy, P.; Lilenbaum, W. First isolation and characterization of *Leptospira interrogans* serogroup Australis from swine in Brazil. *Pesqui. Veterinária Bras.* **2015**, *35*, 6–8. [CrossRef]
37. Scanziani, E.; Giusti, A.M.; Tagliabue, S.; Luini, M.; Conti, G.; Calzolai, C.; Galli, L. Indagine sulla diffusione della leptospirosi in suini da macello. *Arch. Vet. Ital.* **1993**, *44*, 161–165.
38. Tagliabue, S.; Farina, R. Inchiesta sieroepidemiologica sulla diffusione della leptospirosi ta gli animali domestici ed alcune specie selvatiche. *Sel. Vet.* **1995**, *36*, 11–12.
39. Tagliabue, S.; Raffo, A.; Foni, E.; Candotti, R.; Barigazzi, G. Anticorpi per *Leptospira interrogans* in sieri di cinghiale selvatico (*Sus scrofa*) nell'Appennino parmense. In Proceedings of the Convegno Nazionale di Ecopatologia della fauna selvatica, Bologna, Italy, 9–11 February 1995.
40. Ellis, W.A.; Thiermann, A.B. Isolation of *Leptospira interrogans* serovar bratislava from sows in Iowa. *Am. J. Vet. Res.* **1986**, *47*, 1458–1460.

41. Hartman, E.G.; Brummelman, B.; Dikken, H. Leptospirae of serotype lora of the serogroup Australis isolated for the first time from swine in the Netherlands. *Tijdschr. Diergeneeskd.* **1975**, *100*, 421–425.
42. Ellis, W.A.; McParland, P.J.; Bryson, D.G.; Cassells, J.A. Prevalence of Leptospira infection in aborted pigs in Northern Ireland. *Vet. Rec.* **1986**, *118*, 63–65. [CrossRef]
43. Ellis, W.A. Leptospira australis infection in pigs. *Pig Vet. J.* **1989**, *22*, 83–92.
44. Schönberg, A.; Hahn-Hey, B.; Kämpe, U.; Schmidt, K.; Ellis, W.A. The isolation and identification of *Leptospira interrogans* serovar bratislava from a pig in Germany. *Zentralbl. Veterinarmed. B* **1992**, *39*, 362–368. [CrossRef] [PubMed]
45. Agunloye, C.A. Agglutinating antibodies to leptospires in slaughters pigs in Ibadan, Nigeria. *Trop. Vet.* **2001**, *19*, 188–190.
46. Kikuchi, N.; Shikano, M.; Hatanaka, M.; Takahashi, T.; Mori, K.; Fujii, T.; Furuya, T. Prevalence of leptospiral antibody in sows in Japan. *J. Vet. Epidemiol.* **2013**, *13*, 95–99. [CrossRef]
47. Choi, C.; Park, Y.C.; Park, M.A.; Yoo, C.K.; Park, M.Y.; Chae, C. Seroprevalence of Leptospira serovars in Korean sows. *Vet. Rec.* **2001**, *148*, 416. [CrossRef]
48. Ellis, W.A.; Montgomery, J.M.; Thiermann, A.B. Restriction endonuclease analysis as a taxonomic tool in the study of pig isolates belonging to the Australis serogroup of *Leptospira interrogans*. *J. Clin. Microbiol.* **1991**, *29*, 957–961. [CrossRef]
49. Bolin, C.A.; Cassels, J.A. Isolation of *Leptospira interrogans* server bratislava and hard from swine at slaughter. *J. Vet. Diagnostic Investig.* **1992**, *4*, 87–89. [CrossRef]
50. Bolin, C.A.; Cassels, J.A. Isolation of *Leptospira interrogans* server bratislava from still born and weak pigs in Iowa. *J. Am. Vet. Med. Assoc.* **1990**, *196*, 1601–1604.
51. Boqvist, S.; Montgomery, J.M.; Hurst, M.; Thu, H.T.V.; Engvall, E.O.; Gunnarsson, A.; Magnusson, U. Leptospira in slaughtered fattening pigs in southern Vietnam: Presence of the bacteria in the kidneys and association with morphological findings. *Vet. Microbiol.* **2003**, *93*, 361–368. [CrossRef]
52. Ellis, W.A.; McParland, P.J.; Bryson, D.G.; Cassel, J.A. Boars as carriers of leptospires of the Australis serogroup on farms with an abortion problem. *Vet. Rec.* **1989**, *118*, 563. [CrossRef]
53. Poonacha, K.B.; Smith, B.J.; Donahue, J.M.; Tramontin, R.R.; Tuttle, P.A.; Hong, C.B.; Giles, R.C. Leptospiral abortion in horses in central Kentucky. In Proceedings of the Annual Convention of the American Association of Equine Practitioners, Lexington, KY, USA, 2–5 December 1990; pp. 397–402.
54. Miraglia, F.; Moreno, A.M.; Gomes, C.R.; Paixão, R.; Liuson, E.; Morais, Z.M.; Maiorka, P.; Seixas, F.K.; Dellagostin, O.A.; Vasconcellos, S.A. Isolation and characterization of *Leptospira interrogans* from pigs slaughtered in São Paulo State, Brazil. *Braz. J. Microbiol.* **2008**, *39*, 501–507. [CrossRef]
55. Christmas, B.W.; Bragger, J.M.; Till, D.G. Dairy farm fever in New Zealand: Isolation of L pomona and L hardjo from a local outbreak. *New Zealand Med. J.* **1974**, *79*, 904–906. [PubMed]
56. Firinu, A.; Ponti, M.N.; Patta, C.; Oggiano, A.; Ruiu, A.; Cabras, P.; Maestrale, C.; Cossu, P.; Pintore, A. Serologic survey on some transmissible diseases among wild boars and free ranging pigs in Sardinia. In *Tradition and Innovation in Mediterranean Pig Production, Proceedings of the International Symposium on Mediterranean Pig, Evora, Portugal, 26–28 November 1998*; Almeida, J.A., Tirapicos, N.J., Eds.; CIHEAM: Zaragoza, Spain, 2000; pp. 309–312.
57. Żmudzki, J.; Jabłoński, A.; Nowak, A.; Zębek, S.; Arent, Z.; Bocian, Ł.; Pejsak, Z. First overall report of Leptospira infections in wild boars in Poland. *Acta Vet. Scand.* **2016**, *58*, 3. [CrossRef] [PubMed]
58. Ellis, W.A. Animal leptospirosis. *Curr. Top. Microbiol. Immunol.* **2015**, *387*, 99. [PubMed]
59. Schollum, L.M.; Blackmore, D.K. Leptospirosis of pig farmers: The results of a serological survey. *New Zealand Med. J.* **1982**, *95*, 299–301. [PubMed]
60. Tagliabue, S. Nuove acquisizioni nella diagnosi di leptospirosi. In Proceedings of the XLIV Convegno Nazionale della Società Italiana delle Scienze Veterinarie, Stresa, Italy, 27–29 September 1990.
61. Tagliabue, S.; Figarolli, B.M.; D'Incau, M.; Foschi, G.; Gennero, M.S.; Giordani, R.; Natale, A.; Papa, P.; Ponti, N.; Scaltrito, D.; et al. Serological surveillance of Leptospirosis in Italy: Two-year national data (2010–2011). *Vet. Ital.* **2016**, *52*, 129–138.
62. Delbem, Á.C.B.; de Freitas, J.C.; Bracarense, A.P.F.R.L.; Müller, E.E.; de Oliveira, R.C. Leptospirosis in slaughtered sows: Serological and histopathological investigation. *Braz. J. Microbiol.* **2002**, *33*, 174–177. [CrossRef]

63. Scanziani, E.; Crippa, L.; Giusti, A.M.; Luini, M.; Pacciarini, M.L.; Tagliabue, S.; Cavalletti, E. *Leptospira interrogans* serovar sejroe infection in a group of laboratory dogs. *Lab. Anim.* **1995**, *29*, 300–306. [CrossRef]
64. Paz-Soldán, S.V.; Dianderas, M.T.; Windsor, R.S. *Leptospira interrogans* serovar canicola: A causal agent of sow abortions in Arequipa, Peru. *Trop. Anim. Health Prod.* **1991**, *23*, 233–240. [CrossRef]
65. Rühl-Fehlert, C.I.; Brem, S.; Feller, W.; Kopp, H.; Meyer, P.; Rinke, M. Clinical, microbiological and pathological observations in laboratory beagle dogs infected with leptospires of the serogroup sejroe. *Exp. Toxicol. Pathol.* **2000**, *52*, 201–207. [CrossRef]
66. Collares-Pereira, M.; Korver, H.; Thi, B.V.C.; Santos-Reis, M.; Bellenger, E.; Baranton, G.; Terpstra, W.J. Analysis of Leptospira isolates from mainland Portugal and the Azores islands. *FEMS Microbiol. Lett.* **2000**, *185*, 181–187. [CrossRef]
67. das Neves Paiva-Cardoso, M.; Arent, Z.; Gilmore, C.; Hartskeerl, R.; Ellis, W.A. Altodouro, a new Leptospira serovar of the Pomona serogroup isolated from rodents in northern Portugal. *Infect. Genet. Evol. J. Mol. Epidemiol. Evol. Genet. Infect. Dis.* **2013**, *13*, 211–217. [CrossRef] [PubMed]
68. Rocha, T.; Ellis, W.A.; Montgomery, J.; Gilmore, C.; Regalla, J.; Brem, S. Microbiological and serological study of leptospirosis in horses at slaughter: First isolations. *Res. Vet. Sci.* **2004**, *76*, 199–202. [CrossRef] [PubMed]
69. Rocha, T. Isolation of *Leptospira interrogans* serovar mozdok from aborted swine fetuses in Portugal. *Vet. Rec.* **1990**, *126*, 602. [PubMed]
70. Majetic, Z.S.; Galloway, R.; Sabljic, E.R.; Milas, Z.; Perko, V.M.; Habus, J.; Margaletic, J.; Pernar, R.; Turk, N. Epizootiological survey of small mammals as Leptospira spp. reservoirs in Eastern Croatia. *Acta Trop.* **2014**, *131*, 111–116. [CrossRef] [PubMed]
71. Ellis, W.A. Control of canine leptospirosis in Europe: Time for a change? *Vet. Rec.* **2010**, *167*, 602–605. [CrossRef]
72. Renaud, C.; Andrews, S.; Djelouadji, Z.; Lecheval, S.; Corrao-Revol, N.; Buff, S.; Demont, P.; Kodjo, A. Prevalence of the Leptospira serovars bratislava, grippotyphosa, mozdok and pomona in French dogs. *Vet. J.* **2013**, *196*, 126–127. [CrossRef]
73. Pritchard, D.G.; Todd, N.; Barlow, A.; Little, S.A. Outbreak of *Leptospira interrogans* serovar mozdok in sows in Dorset, England. *Isr. J. Vet. Med.* **1987**, *43*, 343.
74. Barlow, A.M. Reproductive failure in sows associated with Leptospira Mozdok from a wildlife source. *Pig J.* **2004**, *54*, 123–131.
75. Hathaway, S.C.; Todd, J.N.; Headlam, S.A.; Jeffrey, M. Possible role of leptospires of the Pomona serogroup in sporadic bovine abortion in the south west of England. *Vet. Rec.* **1984**, *115*, 623–626. [CrossRef]
76. Da Cunha, C.E.P.; Felix, S.R.; Neto, A.C.P.S.; Campello-Felix, A.; Kremer, F.S.; Monte, L.G.; Amaral, M.G.; Nobre, M.D.O.; Da Silva, É.F.; Hartleben, C.P.; et al. Infection with Leptospira kirschneri Serovar Mozdok: First Report from the Southern Hemisphere. *Am. J. Trop. Med. Hyg.* **2016**, *94*, 519–521. [CrossRef]
77. Jacobs, A.; Harks, F.; Hoeijmakers, M.; Segers, R. A novel octavalent combined Erysipelas, Parvo and Leptospira vaccine provides (cross) protection against infection following challenge of pigs with 9 different *Leptospira interrogans* serovars. *Porc. Heal. Manag.* **2015**, *1*, 16. [CrossRef] [PubMed]
78. Jacobs, A.A.C.; Harks, F.; Hoeijmakers, M.; Collell, M.; Segers, R.P.A.M. Safety and efficacy of a new octavalent combined Erysipelas, Parvo and Leptospira vaccine in gilts against *Leptospira interrogans* serovar Pomona associated disease and foetal death. *Vaccine* **2015**, *33*, 3963–3969. [CrossRef] [PubMed]
79. Ellinghausen, H.C.; Mccullough, W.G. Nutrition of Leptospira Pomona and growth of 13 other serotypes: A serum-free medium employing oleic albumin complex. *Am. J. Vet. Res.* **1965**, *26*, 39–44. [PubMed]
80. Faine, S.; Adler, B.; Bolin, C.; Perolat, B. *Leptospira and Leptospirosis*; MediSci.: Melbourne Vic, Australia, 1999.
81. Faine, S. Guidelines for the control of leptospirosis. In *WHO Offset Publ No. 67*; WHO: Geneva, Switzerland, 1982. Available online: https://apps.who.int/iris/handle/10665/37219 (accessed on 25 April 2020).
82. Savio, M.L.; Pacciarini, M.L.; Cinco, M.; Tagliabue, S. Identification of *Leptospira interrogans* strains by monoclonal antibodies and genomic analysis. *New Microbiol.* **1993**, *16*, 315–321. [PubMed]
83. Terpstra, W.J.; Korver, H.; Schoone, G.J.; von Leeuwen, J.; Schönemann, C.E.; de Jonge-Aglibut, S.; Kolk, A.H. Comparative classification of Leptospira serovars of the Pomona group by monoclonal antibodies and restriction-endonuclease analysis. *Zentralbl. Bakteriol. Mikrobiol. Hyg. A.* **1987**, *266*, 412–421. [CrossRef]
84. Kumar, S.; Stecher, G.; Tamara, K. MEGA7: Molecular Evolutionary Genetics Analysis version 7.0 for bigger datasets. *Mol. Biol. Evol.* **2016**, *33*, 1870–1874. [CrossRef]

85. Bolger, A.M.; Lohse, M.; Usadel, B. Trimmomatic: A flexible trimmer for Illumina sequence data. *Bioinformatics* **2014**, *30*, 2114–2120. [CrossRef]
86. Bankevich, A.; Nurk, S.; Antipov, D.; Gurevich, A.A.; Dvorkin, M.; Kulikov, A.S.; Lesin, V.M.; Nikolenko, S.I.; Pham, S.; Prjibelski, A.D.; et al. SPAdes: A new genome assembly algorithm and its applications to single-cell sequencing. *J. Comput. Biol. A J. Comput. Mol. Cell Biol.* **2012**, *19*, 455–477. [CrossRef]
87. Gurevich, A.; Saveliev, V.; Vyahhi, N.; Tesler, G. QUAST: Quality assessment tool for genome assemblies. *Bioinformatics* **2013**, *29*, 1072–1075. [CrossRef]
88. Menzel, P.; Ng, K.L.; Krogh, A. Fast and sensitive taxonomic classification for metagenomics with Kaiju. *Nat. Commun.* **2016**, *7*, 11257. [CrossRef]

© 2020 by the authors. Licensee MDPI, Basel, Switzerland. This article is an open access article distributed under the terms and conditions of the Creative Commons Attribution (CC BY) license (http://creativecommons.org/licenses/by/4.0/).

pathogens

MDPI

Article

Identification of Serogroups Australis and Icterohaemorrhagiae in Two Dogs with a Severe Form of Acute Leptospirosis in Italy

Andrea Balboni [1], Silvia Zamagni [1], Cristina Bertasio [2], Maria Beatrice Boniotti [2], Roberta Troìa [1], Mara Battilani [1,*] and Francesco Dondi [1]

1 Department of Veterinary Medical Sciences, Alma Mater Studiorum, University of Bologna, 40064 Ozzano dell'Emilia, Bologna, Italy; a.balboni@unibo.it (A.B.); silvia.zamagni5@unibo.it (S.Z.); roberta.troia2@unibo.it (R.T.); f.dondi@unibo.it (F.D.)

2 National Reference Centre for Animal Leptospirosis (NRCL), Istituto Zooprofilattico Sperimentale della Lombardia ed Emilia Romagna "Bruno Ubertini", via Bianchi 7/9, 25121 Brescia, Italy; cristina.bertasio@izsler.it (C.B.); mariabeatrice.boniotti@izsler.it (M.B.B.)

* Correspondence: mara.battilani@unibo.it

Received: 17 April 2020; Accepted: 4 May 2020; Published: 6 May 2020

check for updates

Abstract: Leptospirosis is an infectious disease that causes serious illness in dogs. For this reason, epidemiological and clinical studies focusing on disease characterization are widely advocated. The aim of this study was to characterize the leptospires identified in dogs with confirmed symptomatic acute leptospirosis. *Leptospira* spp. DNA detected in urine, blood, or both samples from nine infected dogs was analyzed using the multi-locus sequence typing (MLST) technique. Leptospires from two dogs were successfully typed: one was identified as belonging to Sequence Type (ST) 17 and one to ST198, both within the *L. interrogans* species, serogroups Icterohaemorrhagiae and Australis, respectively. Based on the results of routine serologic tests, antibodies reactive toward these serogroups are commonly revealed in dogs in Italy. This study provides the first molecular analysis that identifies infecting *Leptospira* directly on DNA from biological samples of dogs, showing that serogroup Australis can lead to a severe clinical presentation of leptospirosis in infected dogs.

Keywords: Australis; canine leptospirosis; Icterohaemorrhagiae; multi-locus sequence typing

1. Introduction

Leptospirosis is a worldwide zoonotic disease affecting many mammalian species [1]. Leptospiral infection in dogs causes severe clinical manifestations, which can lead to death in 28–70% of cases [2,3]. In Europe, the serological prevalence of leptospirosis in dogs is about 25% [4,5]. A similar prevalence (29.9%) is reported in Italy according to a study performed in the whole Italian area [6], while the prevalence reported in a more recent paper in North-Central Italy was 8.65% [7]. Despite the high serological prevalence of the infection, data obtained by molecular typing techniques to identify *Leptospira* strains are lacking in our country.

The diagnosis of canine leptospirosis is frequently carried out with serological tests such as the microagglutination test (MAT) performed upon admission or in paired serum samples, as previously recommended [8]. According to the available literature, MAT on serum samples is not able to appropriately recognize the infecting serovar, being able to only identify the serogroup. The serogroup with the highest MAT titer is generally considered the infecting one [9]. This MAT interpretation, however, can lead to flawed conclusions: due to the presence of common antigens among serogroups and the potential for in vitro cross-reaction, several serogroups with high antibody titers can sometimes be displayed by the same dog [10]. Moreover, the serogroup with the highest MAT titer frequently changes over time and between different laboratories, suggesting that it might not really represent the infecting serogroup [10].

PCR or real-time PCR (qPCR) is often carried out to detect *Leptospira* spp. DNA in a variety of biological samples and to diagnose leptospirosis in dogs. Although molecular tests showed low diagnostic sensitivity as they might give a large number of false negative results [11], different techniques have been adopted for the typing of *Leptospira* strains by analyzing the bacterial genome or its specific regions [12–15]. In particular, the multi-locus sequence typing (MLST) technique is able to characterize the genetic profile of *Leptospira* strains by sequencing and analyzing specific fragments of some bacterial house-keeping genes, thus identifying specific sequence types (STs).

The aim of this study was to characterize by MLST analysis the DNA of leptospires detected in dogs affected by acute leptospirosis.

2. Results

Blood and urine samples from nine dogs with acute leptospirosis and with a positive qPCR were used for the study. These dogs were part of a previous study on leptospirosis conducted by our research group [11]. Six out of nine dogs were intact males, 2/9 were spayed females, and 1/9 was an intact female. The median age was four years (range 1–12). Three out of nine dogs were mixed-breed; 2/9 were Labrador retriever; and the remaining 4/9 were Jack Russell terrier, German Shepherd, Weimaraner, Leonberger, and Kurzhaar. Seven out nine dogs had an outdoor lifestyle, whereas only 2/9 had an urban lifestyle. Five out of nine dogs had been vaccinated with a bivalent vaccine, while 4/9 had not been correctly vaccinated. Finally, 4/9 dogs survived, while 5/9 died or were humanely euthanized. The MLST analysis was performed using the scheme proposed by Boonsilp and colleagues [16] on the DNA extracted from the included samples. A complete MLST profile was obtained from a blood sample and from a urine sample belonging to two dogs, Case 1 and Case 2, respectively (GenBank ID: MT411548-MT411561). In Case 1, the infecting Leptospira belonged to ST17, while in Case 2, the genotype of the infecting Leptospira was ST198 (Figure 1). We were unable to achieve a successful PCR amplification in MLST loci in the samples from the remaining seven dogs, probably due to the low amount of leptospiral DNA present.

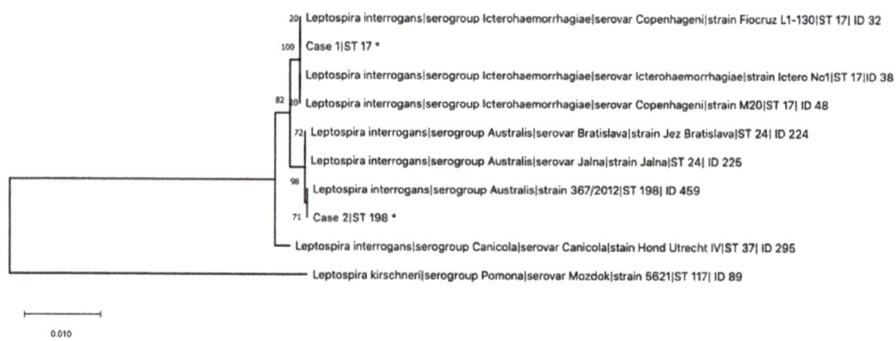

Figure 1. Maximum likelihood tree built on concatenated sequences of the seven multi-locus sequence typing (MLST) loci (3111 bp) of the scheme proposed by Boonsilp and colleagues [16]. Phylogeny was conducted in MEGA X using the Tamura–Nei model, and bootstrap values are indicated on the respective branches. Additional sequences included in the alignment were retrieved from the Pubmlst database. * indicates the leptospires genotyped in this study from dogs of Case 1 (Sequence Type 17 (ST17)) and Case 2 (ST198).

Leptospira from Case 1 clustered with international strains characterized as *L. interrogans* serogroup Icterohaemorrhagiae, serovars Icterohaemorrhagiae and Copenhageni, while *Leptospira* from Case 2 clustered with strain 367/2012, reported as *L. interrogans* serogroup Australis (Figure 1). As depicted in Figure 1, ST198 resulted in being very similar to ST24, found in *L. interrogans* Jalna and Bratislava, having all identical alleles except for the pntA gene, which showed the substitution 234C-T and was registered in the Pubmlst database as "allele 66".

Both Case 1 and Case 2 had a severe and acute clinical presentation of the disease with acute kidney injury (AKI), systemic inflammation, and evidence of damage/dysfunction of at least one organ other than the kidneys. Case 1 survived and was discharged from the hospital after ten days of hospitalization. This was a one-year old, intact male Weimaraner. MAT was negative at the time of admission, but a positive convalescent MAT was evaluated after seven days from presentation (1:400 serogroup Icterohaemorrhagiae, 1:200 serogroup Canicola). Initial diagnosis was carried out with qPCR from blood collected upon admission, whereas the concurrent urine sample was negative. Case 2 died 24 h after hospitalization. This dog was a twelve-year old, spayed female Labrador retriever. The patient was negative for MAT upon admission, and no convalescent test was performed due to the early death of the patient. The qPCR was positive both for blood and urine samples. Neither of these two cases had been properly vaccinated, as previously recommended [17], and both had an outdoor lifestyle, with free access to swampy areas and, potentially, to wild reservoirs of leptospirosis.

3. Discussion

In this study, we performed the molecular characterization of the *Leptospira* infecting two dogs with a confirmed diagnosis of acute leptospirosis. In Case 1, the infecting *Leptospira* belonged to ST17, related to *L. interrogans* serogroup Icterohaemorrhagiae, while Case 2 resulted in being affected by *Leptospira* ST198 related to *L. interrogans* serogroup Australis. Based on the MAT results, it is quite common to reveal antibodies reactive toward these serogroups in dogs in Italy [6,7]. Attempts to deduce the infecting serogroup from serological data were made, but they were subjected to several limitations, as previously stated. Hence, tests able to type *Leptospira* strains correctly would allow better detailing the epidemiology of the disease in different geographic areas, to target the need for specific vaccines, as well as to highlight potential serogroup-specific clinical pictures and prognostic factors in affected animals. This is the first time that the MLST technique has been applied directly to biological samples from dogs in Italy, allowing obtaining *Leptospira* typing before the onset of an IgG response detectable by MAT. In a recent study performed in Brazil, MLST was carried out on two *Leptospira* strains isolated from canine urine samples identifying *L. interrogans* serogroup Icterohaemorrhagiae [12]. A larger study performed in Japan applied the MLST technique on 45 strains isolated from dogs with acute leptospirosis, identifying 16/45 strains belonging to serogroup Australis and 1/45 to serogroup Icterohaemorrhagiae [13]. *L. interrogans* serogroup Australis was lethal in most of those cases (78.6%). Afterwards, the same Japanese working group confirmed these previous data on dogs, identifying 21/75 leptospiral isolates belonging to serogroup Australis, with a mortality rate related to this serogroup as high as 83.3% [14].

In Europe, data regarding the application of MLST in dogs with leptospirosis are lacking. In a recent study, MLST was applied to characterize *L. interrogans* isolated from wild, domestic, and captive animals in Portugal [18]. Serovar Copenhageni (serogroup Icterohaemorrhagiae) was detected in five distinct captive animals and in one rat from Lisbon zoo, and serovar Bratislava (serogroup Australis) was detected in one healthy horse. Another study performed in Northern Ireland demonstrated the presence of *L. interrogans* serogroup Australis in 15/35 (42.7%) horses using MLST [19]. Furthermore, the authors of the cited study compared these results with the frequency of the serogroup Australis in other herbivores (sheep 6.9% and cattle 0.9%).

As is common knowledge, *L. interrogans* serogroup Icterohaemorrhagiae is associated with severe clinical signs, whereas data regarding the pathogenicity of the Australis serogroup and its clinical features during leptospirosis in dogs are limited [3,11,13,14,20,21]. Interestingly, ST198 found in Case 2 was firstly discovered in hedgehogs of Northern Italy by Boniotti and colleagues [22]. Its detection in this study suggests a possible direct transmission between different species or an indirect infection through a contaminated shared environment. Moreover, the current study revealed that ST198 was diffused in Italy and associated with a severe life-threating clinical presentation with multiorgan involvement. A severe clinical presentation in dogs associated with infection with a *Leptospira* ST related to *L. interrogans* serogroup Australis was previously reported in Asia [13,14].

Different types of vaccines are available in Italy: bivalent vaccines containing two serovars (Canicola and Icterohaemorrhagiae), a trivalent vaccine containing serovars Canicola, Icterohaemorrhagiae, and Grippotyphosa, and tetravalent vaccines containing Canicola, Icterohaemorrhagiae, Grippotyphosa, and Bratislava, which belong to the Australis serogroup. Despite the vaccination against serogroup Icterohaemorrhagiae having been used for many years in Italy, this study proves that this serogroup is still circulating and able to infect non-vaccinated dogs. Thus, according to the results reported in this paper, vaccination against both of these serogroups (Icterohaemorrhagiae and Australis) is recommended in Italy.

MLST is currently adopted to correlate leptospiral STs with serotyping; nevertheless, according to the available literature, immunity stimulated by vaccination seems to be serovar-specific [9]. Furthermore, Koizumi end collaborators (2013) suggested that in some cases, MLST is not suitable for identifying serogroups, since it may detect the same ST in more than one serogroup [13]. The same authors compared MLST with multi-locus variable-number tandem repeat analysis (MLVA), concluding that the latter one had a higher discriminatory power and was more concordant with serotyping [14]. Therefore, further studies are needed to develop MLST protocols able to correlate unambiguously STs and serotyping and to identify the serovars belonging to serogroups Australis (serovars Bratislava, Australis, Lora, Muenchen) [23] spread throughout the Italian and European area, in order to target the research on leptospiral vaccines based on circulating serovars.

4. Materials and Methods

4.1. Sample Collection, Inclusion Criteria, and Diagnosis of Leptospirosis

Dogs included in the present study originated from a previous prospective study on canine acute leptospirosis carried out by Troìa and colleagues [11]. In that study, dogs with suspected acute leptospirosis were enrolled on the basis of reported exposure to known risk factors, clinical and clinicopathological signs suggesting AKI, systemic inflammation, or both, and evidence of dysfunction of at least one organ other than the kidney. The diagnosis of leptospirosis was confirmed by a combination of tests: MAT titer at the time of hospital admission (considered positive if the titer was ≥1:800), convalescent MAT titer (fourfold increase from baseline), and/or a positive qPCR from blood, urine samples, or both. Dogs vaccinated for leptospirosis within 15 weeks before hospitalization were excluded from the study [11]. DNA extracted from blood and/or urine samples of nine dogs testing positive for the presence of *Leptospira* spp. DNA using an SYBR Green qPCR was used for the following molecular analyses [11,24]. Four out of 9 dogs had positive blood samples; 4/9 dogs had positive urine samples; and 1/9 dogs had positive blood and urine samples with a *Leptospira* spp. DNA quantity ranging from 1.00×10^0 to 2.09×10^2 DNA copies/µL of extract (median 2.22×10^0) [11].

4.2. Multi-Locus Sequence Typing

The analysis was performed using the seven loci (glmU, pntA, sucA, tpiA, pfkB, mreA, and caiB) scheme proposed by Boonsilp and colleagues [16] following the protocol previously developed for its application directly on DNA extracted from biological samples [25]. Nucleotide sequences for each of the seven genes were trimmed to the reference sizes using the module SeqMan of the Lasergene sequencing analysis software package (DNASTAR, Inc., Madison, WI, USA), and STs were assigned through the Pubmlst database [26]. Nucleotide sequences for each of the seven genes were concatenated and aligned using MegAlign software (Lasergene, DNASTAR). Additional sequences included in the alignment were retrieved from the Pubmlst database, and a phylogenetic tree was conducted in MEGA

X Version 10.1.7 using the maximum likelihood method and the Tamura–Nei model with a bootstrap analysis based on 1000 replicates [27,28].

5. Conclusions

This study used the MLST technique and identified two different *L. interrogans* STs, corresponding to serogroups Icterohaemorrhagiae and Australis, respectively, in two dogs with clinical acute leptospirosis. This was the first time that MLST was applied in Italy to characterize *Leptospira* spp. directly on DNA from biological samples of symptomatic dogs, allowing typing *Leptospira* before the onset of an IgG response detectable by MAT, and independently of the availability of the isolated sample, which requires several days. The MLST technique appeared to be a valuable tool to overcome the limitations of serological testing for epidemiologic analysis of leptospirosis in different countries. In this regard, the results of the current study confirmed that the Australis serogroup can cause a form of acute leptospirosis with a severe clinical picture and supported the development of vaccination protocols with serovars belonging to this serogroup.

Author Contributions: Conceptualization, A.B., M.B. and F.D.; methodology, A.B., S.Z., M.B.B. and F.D.; software, C.B.; formal analysis, A.B., S.Z., C.B., M.B.B. and R.T.; investigation, A.B., S.Z., C.B., M.B.B. and R.T.; resources, M.B.B. and F.D.; data curation, S.Z., C.B. and R.T.; writing—original draft preparation, A.B., S.Z., C.B. and R.T.; writing—review and editing, M.B. and F.D.; supervision, A.B. and F.D.; All authors have read and agreed to the published version of the manuscript.

Funding: This research received no external funding.

Conflicts of Interest: The authors declare no conflict of interest.

References

1. Bharti, A.R.; Nally, J.E.; Ricaldi, J.N.; Matthias, M.A.; Diaz, M.M.; Lovett, M.A.; Levett, P.N.; Gilman, R.H.; Willig, M.R.; Gotuzzo, E.; et al. Leptospirosis: A zoonotic disease of global importance. *Lancet Infect. Dis.* **2003**, *3*, 757–771. [CrossRef]
2. Goldstein, R.E.; Lin, R.C.; Langston, C.E.; Scrivani, P.V.; Erb, H.N.; Barr, S.C. Influence of infecting serogroup on clinical features of leptospirosis in dogs. *J. Vet. Intern. Med.* **2006**, *20*, 489–494. [CrossRef] [PubMed]
3. Kohn, B.; Steinicke, K.; Arndt, G.; Gruber, A.D.; Guerra, B.; Jansen, A.; Kaser-Hotz, B.; Klopfleisch, R.; Lotz, F.; Luge, E.; et al. Pulmonary abnormalities in dogs with leptospirosis. *J. Vet. Intern. Med.* **2010**, *24*, 1277–1282. [CrossRef] [PubMed]
4. Delaude, A.; Rodriguez-Campos, S.; Dreyfus, A.; Counotte, M.J.; Francey, T.; Schweighauser, A.; Lettry, S.; Schuller, S. Canine leptospirosis in Switzerland-A prospective cross-sectional study examining seroprevalence, risk factors and urinary shedding of pathogenic leptospires. *Prev. Vet. Med.* **2017**, *141*, 48–60. [CrossRef]
5. López, M.C.; Vila, A.; Rodón, J.; Roura, X. Leptospira seroprevalence in owned dogs from Spain. *Heliyon* **2019**, *5*, 02373. [CrossRef]
6. Tagliabue, S.; Figarolli, B.M.; D'Incau, M.; Foschi, G.; Gennero, M.S.; Giordani, R.; Natale, A.; Papa, P.; Ponti, N.; Scaltrito, D.; et al. Serological surveillance of Leptospirosis in Italy: Two-year national data (2010–2011). *Vet. Ital.* **2016**, *52*, 129–138.
7. Bertelloni, F.; Cilia, G.; Turchi, B.; Pinzauti, P.; Cerri, D.; Fratini, F. Epidemiology of leptospirosis in North-Central Italy: Fifteen years of serological data (2002–2016). *Comp. Immunol. Microbiol. Infect. Dis.* **2019**, *65*, 14–22. [CrossRef]
8. Schuller, S.; Francey, T.; Hartmann, K.; Hugonnard, M.; Kohn, B.; Nally, J.E.; Sykes, J. European consensus statement on leptospirosis in dogs and cats. *J. Small Anim. Pract.* **2015**, *56*, 159–179. [CrossRef]
9. Levett, P.N. Leptospirosis. *Clin. Microbiol. Rev.* **2001**, *14*, 296–326. [CrossRef]
10. Miller, M.D.; Annis, K.M.; Lappin, M.R.; Lunn, K.F. Variability in results of the microscopic agglutination test in dogs with clinical leptospirosis and dogs vaccinated against leptospirosis. *J. Vet. Intern. Med.* **2006**, *25*, 426–432. [CrossRef]
11. Troìa, R.; Balboni, A.; Zamagni, S.; Frigo, S.; Magna, L.; Perissinotto, L.; Battilani, M.; Dondi, F. Prospective evaluation of rapid point-of-care tests for the diagnosis of acute leptospirosis in dogs. *Vet. J.* **2018**, *237*, 37–42. [CrossRef] [PubMed]

12. Miotto, B.A.; Tozzi, B.F.; Penteado, M.S.; Guilloux, A.G.A.; Moreno, L.Z.; Heinemann, M.B.; Moreno, A.M.; Lilenbaum, W.; Hagiwara, M.K. Diagnosis of acute canine leptospirosis using multiple laboratory tests and characterization of the isolated strains. *BMC Vet. Res.* **2018**, *14*, 222. [CrossRef] [PubMed]
13. Koizumi, N.; Muto, M.M.; Akachi, S.; Okano, S.; Yamamoto, S.; Horikawa, K.; Harada, S.; Funatsumaru, S.; Ohnishi, M. Molecular and serological investigation of Leptospira and leptospirosis in dogs in Japan. *J. Med. Microbiol.* **2013**, *62*, 630–636. [CrossRef] [PubMed]
14. Koizumi, N.; Muto, M.M.; Izumiya, H.; Suzuki, M.; Ohnishi, M. Multiple-locus variable-number tandem repeat analysis and clinical characterization of Leptospira interrogans canine isolates. *J. Med. Microbiol.* **2015**, *64*, 288–294. [CrossRef] [PubMed]
15. Larson, C.R.; Dennis, M.; Nair, R.V.; Llanes, A.; Peda, A.; Welcome, S.; Rajeev, S. Isolation and characterization of Leptospira interrogans serovar Copenhageni from a dog from Saint Kitts. *JMM Case Rep.* **2017**, *4*, e005120. [CrossRef]
16. Boonsilp, S.; Thaipadungpanit, J.; Amornchai, P.; Wuthiekanun, V.; Bailey, M.S.; Holden, M.T.; Zhang, C.; Jiang, X.; Koizumi, N.; Taylor, K.; et al. A single multilocus sequence typing (MLST) scheme for seven pathogenic Leptospira species. *PLoS Negl. Trop Dis.* **2013**, *7*, e1954. [CrossRef]
17. Day, M.J.; Horzinek, M.C.; Schultz, R.D.; Squires, R.A. Vaccination Guidelines Group (VGG) of the World Small Animal Veterinary Association (WSAVA). WSAVA Guidelines for the vaccination of dogs and cats. *J. Small Anim. Pract.* **2016**, *57*, E1–E45. [CrossRef]
18. Ferreira, A.S.; Ahmed, A.; Rocha, T.; Vieira, M.L.; Paiva-Cardoso, M.D.N.; Mesquita, J.R.; van der Linden, H.; Goris, M.; Thompson, G.; Hartskeerl, R.A.; et al. Genetic diversity of pathogenic leptospires from wild, domestic and captive host species in Portugal. *Transbound. Emerg. Dis.* **2020**, *67*, 852–864. [CrossRef]
19. Arent, Z.; Gilmore, C.; Brem, S.; Ellis, W.A. Molecular studies on European equine isolates of Leptospira interrogans serovars Bratislava and Muenchen. *Infect. Genet. Evol.* **2015**, *34*, 26–31. [CrossRef]
20. Renaud, C.; Andrews, S.; Djelouadji, Z.; Lecheval, S.; Corrao-Revol, N.; Buff, S.; Demont, P.; Kodjo, A. Prevalence of the Leptospira serovars bratislava, grippotyphosa, mozdok and pomona in French dogs. *Vet. J.* **2013**, *196*, 126–127. [CrossRef]
21. Mastrorilli, C.; Dondi, F.; Agnoli, C.; Turba, M.E.; Vezzali, E.; Gentilini, F. Clinicopathologic features and outcome predictors of Leptospira interrogans Australis serogroup infection in dogs: A retrospective study of 20 cases (2001–2004). *J. Vet. Intern. Med.* **2007**, *21*, 3–10. [CrossRef]
22. Boniotti, M.B.; Gelmini, L.; Carra, E.; Figarolli, B.M.; D'incau, M.; Tagliabue, S. Leptospira Interrogans Serogroup Australis In Hedgehog In Northern Italy. In Proceedings of the International Leptospirsis Society, Fukuoka, Japan, 7–11 October 2013.
23. Nielsen, J.N.; Armstrong, C.H.; Nielsen, N.C. Relationship among selected Leptospira interrogans serogroups as determined by nucleic acid hybridization. *J. Clin. Microbiol.* **1989**, *27*, 2724–2729. [CrossRef] [PubMed]
24. Stoddard, R.A.; Gee, J.E.; Wilkins, P.P.; McCaustland, K.; Hoffmaster, A.R. Detection of pathogenic Leptospira spp. through TaqMan polymerase chain reaction targeting the LipL32 gene. *Diagn. Microbiol. Infect. Dis.* **2009**, *64*, 247–255. [CrossRef] [PubMed]
25. Weiss, S.; Menezes, A.; Woods, K.; Chanthongthip, A.; Dittrich, S.; Opoku-Boateng, A.; Kimuli, M.; Chalker, V. An Extended Multilocus Sequence Typing (MLST) Scheme for Rapid Direct Typing of Leptospira from Clinical Samples. *PLoS Negl. Trop. Dis.* **2016**, *10*, e0004996. [CrossRef] [PubMed]
26. Leptospira spp. MLST Database. Available online: https://pubmlst.org/leptospira/ (accessed on 9 April 2020).
27. Tamura, K.; Nei, M. Estimation of the number of nucleotide substitutions in the control region of mitochondrial DNA in humans and chimpanzees. *Mol. Biol. Evol.* **1993**, *10*, 512–526. [PubMed]
28. Kumar, S.; Stecher, G.; Li, M.; Knyaz, C.; Tamura, K. MEGA X: Molecular Evolutionary Genetics Analysis across Computing Platforms. *Mol. Biol. Evol.* **2018**, *35*, 1547–1549. [CrossRef]

© 2020 by the authors. Licensee MDPI, Basel, Switzerland. This article is an open access article distributed under the terms and conditions of the Creative Commons Attribution (CC BY) license (http://creativecommons.org/licenses/by/4.0/).

pathogens

MDPI

Article

Detection of New *Leptospira* Genotypes Infecting Symptomatic Dogs: Is a New Vaccine Formulation Needed?

Cristina Bertasio [1,*], Maria Beatrice Boniotti [1], Laura Lucchese [2], Letizia Ceglie [2], Laura Bellinati [2], Matteo Mazzucato [2], Tommaso Furlanello [3], Mario D'Incau [1] and Alda Natale [2]

1 National Reference Centre for Animal Leptospirosis, Istituto Zooprofilattico Sperimentale della Lombardia ed Emilia Romagna "Bruno Ubertini", 25121 Brescia, Italy; mariabeatrice.boniotti@izsler.it (M.B.B.); mario.dincau@izsler.it (M.D.)

2 Istituto Zooprofilattico Sperimentale delle Venezie, Viale dell'Università, 35020 Legnaro, Italy; llucchese@izsvenezie.it (L.L.); lceglie@izsvenezie.it (L.C.); lbellinati@izsvenezie.it (L.B.); mmazzucato@izsvenezie.it (M.M.); anatale@izsvenezie.it (A.N.)

3 "San Marco" Veterinary Clinica and Laboratory, Via dell' Industria, 35030 Veggiano, Italy; tf@sanmarcovet.it

* Correspondence: cristina.bertasio@izsler.it; Tel.: +39-030-229-0209

Received: 28 May 2020; Accepted: 17 June 2020; Published: 18 June 2020

check for updates

Abstract: Leptospirosis in dogs has been largely described worldwide, and epidemiological studies have been mainly based on serological data. This study aims to detect and genotype leptospires affecting symptomatic dogs in Northeast Italy between 2013 and 2019. Overall, 1631 dogs were tested using real-time PCR, and leptospires from 193 dogs were subjected to Multilocus Sequence Typing and a Multiple Loci Variable-number Tandem Repeat Analysis. Leptospires were successfully isolated from 15 symptomatic dogs. Six distinct Sequence Types (STs) were found for 135 leptospires, with 3 STs characterizing *Leptospira interrogans* (ST17, ST198 and ST24), 2 STs characterizing *Leptospira kirschneri* (ST117 and ST289) and 1 ST characterizing *Leptospira borgpetersenii* (ST155), revealing the circulation of the serogroups Icterohaemorrhagiae, Australis, Sejroe and Pomona. The Multiple Loci Variable-number Tandem Repeat Analysis of 17 samples did not result in any additional discrimination. Genotypes were compared with those of strains present in the historical internal database, and possible transmission chains were identified from rat, mouse, hedgehog and pig. This work highlights the importance of molecular methods in revealing and identifying circulating *Leptospira* strains, and it also encourages the evaluation of the ability of commercially available vaccines to reduce the disease burden among dogs.

Keywords: leptospirosis; dog; real-time PCR; genotyping; epidemiology; multilocus sequence typing; multiple loci variable-number tandem repeat analysis

1. Introduction

Leptospirosis is a widespread zoonotic disease caused by infection with a pathogenic species of the genus Leptospira [1]. It is a worldwide public health and veterinary problem involving many domestic and wild animal species. Dogs have been known to be hosts of pathogenic leptospires since 1931, when Klarenbeek and Schuffner first isolated leptospires from the urine of a dog affected by nephritis [2]. Dogs are highly susceptible to infection, and they act as a sentinel species for the environmental risk to humans because of their high level of environmental exposure to pathogenic leptospires. Canine leptospirosis has been largely described worldwide [1–3], and its seroprevalence varies in accordance with geographic location, as follows: 1.8% in Australia [4], 7.3% in China [5], 9.9% in Brazil [6], 14.63% in Iran [7], 17.1% in the USA [8], 25.8% in Spain [9] and 71.1% in India [10]. In Italy, a 2002 survey conducted on kenneled dogs reported a seroprevalence of 29.4% [11], and national data collected in 2010–2011, involving more than 3000 dogs, described a seropositivity of 29.9% [12]. Nevertheless, these studies were not standardized, and some were partially biased by vaccine responses, the panel of serovars adopted for the Micro Agglutination Test (MAT) and sample collection, which was based on clinical suspects. In fact, a recent survey conducted in Italy on both ill and healthy dogs reported a lower seroprevalence of 8% [13].

Canine leptospirosis has mainly been associated with serovars Canicola and Icterohaemorrhagiae, two serogroups included within the *Leptospira interrogans* species. In Europe, protective vaccines for dogs against these serovars have been available for ~60 years [14]; however, recently, leptospirosis' epidemiological situation in dogs has changed, and clinical syndromes have been described in association with serovars not included in the traditional vaccines [1,11,15,16]. Because Grippotyphosa and Bratislava have emerged as major causes of canine leptospirosis in Europe [12,17–19], new canine *Leptospira* vaccines containing antigens from up to four different serogroups, Canicola, Icterohaemorrhagiae, Australis and Grippotyphosa, have been introduced in the area [20,21]. In particular, trivalent (serogroups Canicola, Icterohaemorrhagiae and Grippothyphosa) and tetravalent (serogroups Canicola, Icterohaemorrhagiae, Grippothyphosa and Australis) vaccines have been licensed in European countries.

To date, epidemiological studies have been mainly based on serological diagnoses, particularly MAT, which is based on determining the ability of serial dilutions of the tested serum to agglutinate live leptospiral serovars in vitro. Agglutination is assessed by darkfield microscopy and suggests exposure to a serovar belonging to the corresponding serogroup (but not necessarily to the serovar tested) [22]. It is based on the specific antigenic determinants related to the structural heterogeneity of the lipopolysaccharide, and it is considered the diagnostic test of choice in dogs suspected of having leptospirosis [16,23]. Despite the widespread use of MAT to diagnose leptospirosis in dogs, this assay has limitations linked to its indirect diagnostic nature, such as possible false-negative results for initial infections, cross-reactions and paradoxical reactions early during the course of the disease [22,24,25], the variability of the selected antigenic panel and the subjective interpretation of the results. Furthermore, it generally does not discriminate between vaccinated and infected, increasing the difficulty of interpreting canine tests [26]. In addition, making direct comparisons between different studies is complicated by the variability in cut-off MAT titers used.

Several direct molecular assays, such as real-time PCR targeting various leptospiral genes, have been developed to support serological methods [27–32]. The diagnostic performances of different PCR assays are not equivalent [33], but they have been very useful in confirming diagnoses at the early stages of infection, when antibody titers are at undetectable levels [34], and in testing vaccinated patients because previous vaccination does not interfere with the PCR results [35]. Positive PCR results indicate that leptospiral DNA is present in the sample, but negative blood or urine results do not rule out leptospirosis. In fact, leptospiremia is transient, and urinary shedding is delayed after acute infection and can be intermittent. Furthermore, having received an antibiotic treatment recently may affect the detection of leptospiral DNA [1]. For this reason, PCR results should always be interpreted cautiously and in conjunction with MAT results, and they should take into account the clinical context. Furthermore, although its use is highly feasible, routine diagnostic PCR provides no information on the infecting serovar.

Some methods of molecular typing, such as Multilocus Sequence Typing (MLST) [36] and Multiple Loci Variable-number Tandem Repeat Analysis (MLVA) [37], offer interesting epidemiological perspectives by providing a specific and unique barcode for the infecting *Leptospira* through the analysis of specific fragments of particular bacterial loci. Until recently, the MLST technique was only applicable to isolated strains, because it required relatively large amounts of leptospiral DNA, making its direct use on clinical specimens impossible. This was a considerable limitation because culturing leptospires is challenging, time consuming and requires an equipped laboratory. Moreover, the isolation efficiency is very low and is dependent on the *Leptospira* strain and the use of an antimicrobial treatment prior to collection. Fortunately, a fast and specific method for genotyping *Leptospira* DNA directly from biological samples has been developed [38], allowing the rapid identification of the pathogen without strain isolation and providing the opportunity to investigate all the circulating strains, not just those successfully isolated, which represent a small percentage of the spreading strains. Through the assignment of sequence types (STs), MLST permits objective comparisons between strains of *Leptospira* infecting the same host in different geographic regions or different host species within the same geographic region, providing a helpful and powerful tool in investigating the epidemiology of leptospirosis. Therefore, knowledge of regional epidemiology, which can be reliably assessed only by the identification of locally prevalent strains, is necessary for understanding the infection and transmission chains and for maintaining up-to-date vaccination strategies.

The aims of this study were to provide the existent assessment of the genetic diversity in leptospires infecting dogs in Northeast Italy, highlight the need to update vaccine formulations to improve effective preventive measures and reduce the burden of leptospirosis among the canine population.

2. Results

2.1. Real-Time PCR

Real-time PCR detected pathogenic leptospiral DNA in 264 out of 1631 (16.2%) tested dogs (347 out of 2485 biological samples) (Table 1).

Table 1. Results of real-time PCR performed on canine biological samples.

	Urine	Blood	Kidney	Liver	Lung	Bladder	Spleen	Organ(s)	Not Defined	Total
No. of Positives/Total	162/1095	66/665	75/392	23/115	7/33	2/9	0/6	12/148	0/22	347/2485
Percentage of Positives	14.8%	9.9%	18.6%	20.0%	21.2%	22.2%	0.0%	8.1%	0.0%	14.0%

2.2. Isolation

In total, 15 strains were successfully isolated, all from urine samples. They were identified at serogroup and serovar level using MAT with polyclonal antisera and monoclonal antibodies (mAbs), respectively (Table 2).

Table 2. Isolated samples and their serological typing assessed by the Microscopic Agglutination Test.

No. of Isolates	Species	Serogroup	Serovar
8	*L. interrogans*	Icterohaemorrhagiae	Icterohaemorrhagiae
3	*L. interrogans*	Icterohaemorrhagiae	Copenhageni
2	*L. kirschneri*	Pomona	Mozdok
2	*L. interrogans*	Australis	Australis

2.3. Genotyping Analyzsis

2.3.1. Multilocus Sequence Typing (MLST)

Out of 193 leptospires infecting dogs subjected to a genotyping analysis, the sequence types of 106 were completely defined, 29 were partially defined (4–6 amplifiable loci) and 58 were undefined (0–3 amplifiable loci). Difficulties in amplifying all loci resulted from the low amounts of leptospiral DNA present in the samples. Samples having a partial pattern were considered because they can be still classifiable to unique known STs, which are defined by loci alleles (Table 3).

Table 3. Multilocus Sequence Typing (MLST)-based genotyping using the seven-loci scheme developed by Boonsilp et al. [36].

Genotyped Samples	ST 17	ST 198	ST 24	ST 117	ST 155	ST 289	ST 17 Like	Total
No. of DNA	65 (+17)	8 (+3)	6 (+4)	3 (+2)	4 (+2)	5	(1)	91 (+29)
No. of Isolates	11	2		2				15
Total	76 (+17)	10 (+3)	6 (+4)	5 (+2)	4 (+2)	5	(1)	135
%	68.9%	9.6%	7.4%	5.2%	4.4%	3.7%	0.7%	100.0%

The numbers of samples having complete MLST patterns are reported. Numbers in parentheses indicate the additional samples having a partial pattern. ST: sequence type.

A total of six known distinct STs were detected, with three STs belonging to *L. interrogans* (ST17, ST198 and ST24), two STs belonging to *L. kirschneri* (ST117 and ST289) and one ST belonging to *L. borgpetersenii* (ST155).

The major portion (68.9%) of *Leptospira* infecting dogs was identified as ST17, which identifies *L. interrogans* serogroup Icterohaemorrhagiae (serovar Icterohaemorrhagiae or Copenhageni). Other portions were identified as follows: 9.6% as ST198, which identifies *L. interrogans* serogroup Australis serovar Australis; 7.4% as ST24, which identifies *L. interrogans* serogroup Australis (serovar Bratislava or Jalna); 5.2% as ST117, which identifies *L. kirschneri* serogroup Pomona serovar Mozdok; 4.4% as ST155, which identifies *L. borgpetersenii* serogroup Sejroe and 3.7% as ST289, which identifies *L. kirschneri* serogroup Pomona.

One sample collected from the Emilia-Romagna region in 2019 showed a new genotype (named as ST17-like) that was characterized by a new allelic combination, similar to that of ST17 (*glmU*: 1, *pntA*: 1, *sucA*: 2, *tpiA*: 2, *pfkB*: 10, *mreA*: 4, *caiB*: 8), but having allele 13 instead of 1 for the *glmU* gene. Its *sucA* gene was not amplifiable and, therefore, its MLST profile was incomplete.

Considering an immunity protection of one year after vaccination [39,40], 12 out of 40 dogs that had knowingly been vaccinated were theoretically protected from the clinical disease while 28 were not protected because they had been vaccinated more than 12 months before sample collections, or never vaccinated (Table 4).

Table 4. Vaccination and vital states of sampled dogs and the STs of the infecting *Leptospira*.

ST of Infecting *Leptospira*	Vaccination Status			Vital Status			Total
	Regular	Not Regular	Unknown	Alive	Died	Unknown	
17	3	23	65	5	32	56	91
198	6	1	6	4	1	8	13
24		1	9		1	9	10
117	1	1	5	1	3	3	7
155	1	2	3		1	5	6
289	1		4		1	4	5
17 Like			1			1	1
Total	12	28	95	10	39	86	135

Regular: if the last vaccination occurred less than 12 months before sampling; Not regular: if the vaccination occurred more than 12 months before sampling or never.

Three regularly vaccinated dogs were clinically affected by *L. interrogans* serogroup Icterohaemorrhagiae (ST17), a serogroup included in all the vaccine formulations. The following related anamnestic and prognostic data were available: one dog was a 5-month old puppy that had regularly received the first dose of the vaccine. The dog was developing symptoms but completely recovered; the second was a hunting dog that received a booster dose two months before but with a fatal outcome because it was treated too late, and the third dog commonly frequented a river near its home and was showing the onset of hyperacute symptoms. However, thanks to therapeutic intervention, it completely recovered.

Three regularly vaccinated dogs were infected by serovars belonging to Pomona and Sejroe serogroups (ST117, ST155 and ST289) that were not included in the vaccine formulations. Four out of six vaccinated dogs infected with serovar Australis ST198 had been vaccinated with bivalent or trivalent vaccines that did not contain antigens of the Australis serogroup and one received a tetravalent vaccine, which contained *L. interrogans* Bratislava within Australis serogroup as the antigen.

Overall, for 39 of 135 dogs, leptospirosis was lethal (Table 4). Among them, 32 (82%) were infected by *Leptospira* ST17, 3 (7.7%) by *Leptospira* ST117 and 1 (2.6%) each by *Leptospira* belonging to ST24, ST155, ST198 and ST289 (Table 4). Interestingly, all the STs were able to cause fatal leptospirosis.

2.3.2. Multiple Loci Variable-Number Tandem Repeat Analysis (MLVA)

The MLVA pattern of 15 isolates and 2 DNAs from biological samples were obtained. The failure to amplify some loci was due to two different causes: the absence of the locus for biological reasons in the species under test (for example the locus Lb4 for *L. interrogans* and the loci 10 and Lb4 for *L. kirschneri*) [37] that was indicated as N (negative) or because of intrinsic causes (i.e., low amount of DNA) that do not permit a successful amplification (in this case we used the term "Not Amplifiable"). Twelve samples (11 isolates and 1 DNA) belonging to *L. interrogans* Icterohaemorrhagiae were genotyped as 2-1-7-N-6 for the loci 4-7-10-Lb4-Lb5, respectively. Additionally, two isolates previously typed as ST198 (*L. interrogans* Australis Australis) were genotyped as 2-10-10-N-5, respectively; two isolates with ST117 (*L. kirschneri* Pomona Mozdok) were genotype as 0-1-N-N-3, respectively, and one DNA of *L. interrogans* Australis Jalna/Bratislava (ST24) was genotyped as 4-10-10-N-Not Amplifiable, respectively. The MLVA patterns did not permit any further discrimination of the strains belonging to the same ST, but they corroborated the results obtained by the MAT and/or MLST analysis.

2.4. *Sequencing A Tract of the lic12008 Gene on Strains Belonging to ST17*

A small tract of the *lic12008* genes in 93 samples (both isolates and DNAs) that belonged to ST17 was sequenced for serovar discrimination. In total, 66 gene tracts were successfully sequenced, with 62 identified as serovar Icterohaemorrhagiae and 4 as serovar Copenhageni (Table 5). No product was obtained owing to the low concentrations of leptospiral DNA from 19 isolates, and 8 DNA samples were insufficient for analysis.

Table 5. The *lic12008* gene sequencing of strains previously typed as ST17.

No.	Type	Species	Sequence Type	Serovar Definition	
				By *lic12008* Sequencing	**By MAT**
8	isolate	*L. interrogans*	17	Icterohaemorrhagiae	Icterohaemorrhagiae
2	Isolate	*L. interrogans*	17	Icterohaemorrhagiae	Copenhageni
1	Isolate	*L. interrogans*	17	Copenhageni	Copenhageni
52	DNA	*L. interrogans*	17	Icterohaemorrhagiae	N.A.
3	DNA	*L. interrogans*	17	Copenhageni	N.A.

Serovar identification by *lic12008* sequencing and MAT are compared. N.A.: not applicable.

Interestingly, the serovar identifications by *lic12008* sequencing and MAT were in agreement for nine isolates, but not for two isolates that were identified as serovar Copenhageni using the serological method but as serovar Icterohaemorrhagiae using sequencing.

Two dogs infected with *Leptospira* of serovar Icterohaemorrhagiae were regularly vaccinated: one with a tetravalent vaccine and one with a bivalent vaccine, which both contained antigens of serovar Icterohaemorrhagiae. One dog infected with *Leptospira* serogroup Icterohaemorrhagiae, serovar undefined, was regularly vaccinated with a quadrivalent vaccine containing serovar Icterohaemorrhagiae. In total 18 of 62 dogs infected with *Leptospira* serovar Icterohaemorrhagiae were not optimally protected because they had been vaccinated more than 12 months before sample collection or never. For the four dogs infected with serovar Copenhageni, one had never been vaccinated against *Leptospira* infection, while the vaccination states of the three remaining dogs were unavailable.

2.5. Data Analysis

2.5.1. Geographical Maps Based on Found Genotypes

Figure 1 shows the distribution of the identified STs among the provinces of the five regions of Northeast Italy (Lombardy, Emilia Romagna, Veneto, Trentino Alto Adige and Friuli Venezia Giulia). Both the complete and the partial MLST profiles were reported and placed in a corresponding ST. Two genotyped samples identified as ST17 were excluded from the map because their province could not be determined. The high concentration of genotyped samples in the Veneto region was associated with the large number of samples delivered to the laboratory for *Leptospira*-infection diagnosis owing to the wide involvement of veterinary practitioners in the research, which was less common in the other regions. Despite some variability at the regional level, the STs identified in the tested dogs were found widespread in the investigated area. ST17, the most highly represented ST among the detected genotypes, was present in all five of the regions considered. The limited sampling in some regions may explain why some STs were not detected in all the regions considered.

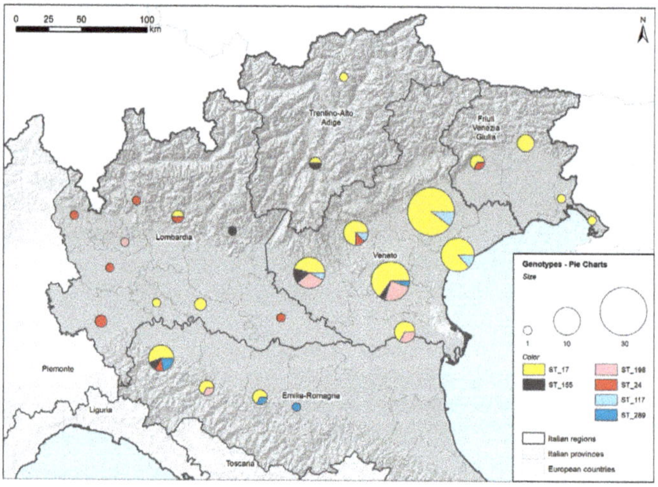

Figure 1. Geographical distribution of Sequence Types (STs). Locations and amounts of genotyped strains, assessed by Multilocus Sequence Typing (MLST), are reported using ST and province in five regions of Northeast Italy (colored in grey) using pie charts. ST17: *L. interrogans* serogroup Icterohaemorrhagiae (serovar Icterohaemorrhagiae or Copenhageni); ST24: *L. interrogans* serogroup Australis (serovar Bratislava or Jalna); ST198: *L. interrogans* serogroup Australis serovar Australis; ST155: *L. borgpetersenii* serogroup Sejroe; ST117: *L. kirschneri* serogroup Pomona serovar Mozdok; ST289: *L. kirschneri* serogroup Pomona.

2.5.2. Phylogenetic Analysis Based on Concatenated MLST Loci

A phylogenetic tree was constructed using the sequences of 3111-bp of concatenated MLST loci (Figure 2). Samples genotyped as ST17 clustered with reference strains *L. interrogans* Copenhageni (strains Wijnberg and M20) and Icterohaemorrhagiae (Bianchi 1 and Ictero I). Leptospires genotyped as ST155 clustered with reference strain *L. borgpetersenii* Sejroe strain Topino 1, and those genotyped as ST24 clustered with reference strains *L. interrogans* Jalna Jalna and Bratislava Riccio 37. Leptospires having ST117 clustered with strain *L. kirschneri* Mozdok 5621, while those having ST198 clustered with strain 367/2012, a recently obtained isolated in Istituto Zooprofilattico Sperimentale della Lombardia e dell'Emilia Romagna laboratory from a hedgehog [41], genotyped as *L. interrogans* serovar Australis and shared six out seven loci with ST24. The phylogenetic analysis indicated that ST289 is closely related to *L. kirschneri* serovar Mozdok, as recently reported [42].

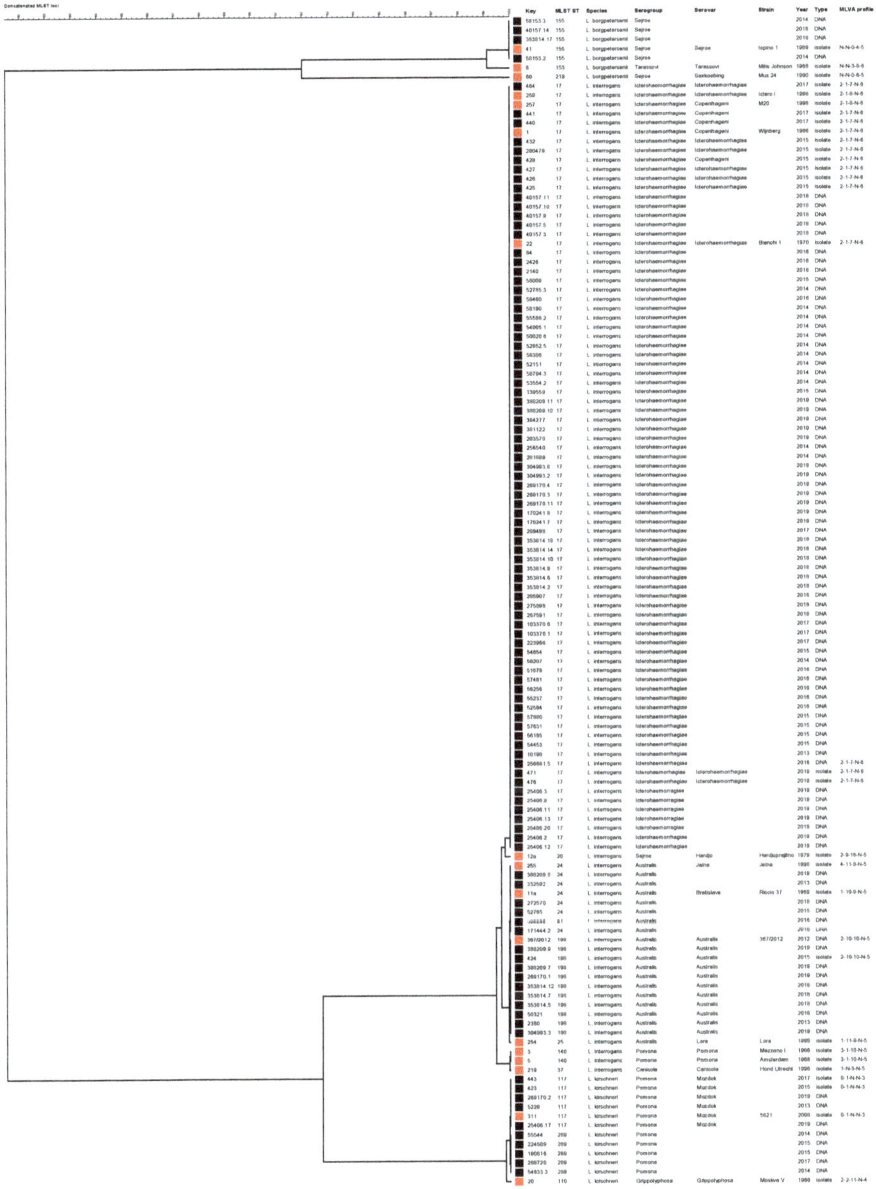

Figure 2. Phylogenetic tree constructed using the 3111-bp concatenated MLST loci. The Unweighted Pair Group Method with Arithmetic mean (UPGMA) was used with a bootstrap analysis based on 1000 replicates. Reference strains are indicated by red squares, while field strains are indicated with black squares. The MLVA pattern of each sample is reported on the right (the order of loci is as follows: VNTR-4, -7, -10, Lb4 and Lb5). N: negative.

2.5.3. Minimum Spanning Tree Based on MLST Data

The genotypes obtained in this study were compared with the STs present in the Istituto Zooprofilattico Sperimentale della Lombardia e dell'Emilia Romagna (IZSLER) database and relative to other host species, using a minimum spanning tree analysis (Figure 3).

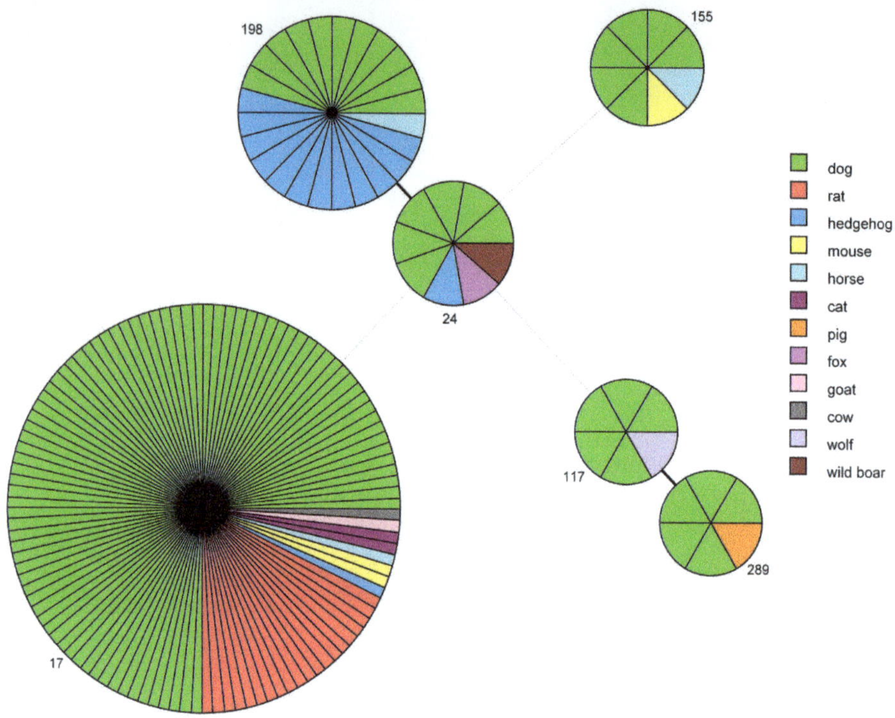

Figure 3. Minimum spanning tree based on sequence types (STs). Each circle represents one specific ST (assigned numbers are located outside the circles), and the number of sectors corresponds to the number of samples having this ST. The tree is color-coded according to host. Solid black lines represent branches between STs that differ at one locus, while dotted light-grey lines represent branches between STs differing at more than four loci.

ST17, which was very frequently identified in dogs, was also found in 20 rats, 2 mice, 2 cats, 1 hedgehog, 1 horse, 1 goat and 1 cow. ST198 was first identified by the IZSLER laboratory in nine hedgehogs collected in 2012 from a collection center at Modena, Italy [41], and it was subsequently found in three more hedgehogs, one dog and one horse. ST24 was identified also in one hedgehog, one wild boar and one fox. ST289, which was first discovered in a pig in Northern Italy in 2014 [42], is a new genotype that has been identified in five dogs. ST117 was also found in a wolf and ST155 was responsible for the infection of one mouse and one horse.

3. Discussion

The absence of available molecular typing data and the complex genomic diversity of strains responsible for canine leptospirosis motivated us to genotype strains from symptomatic dogs in Northeast Italy, sampled between 2013 and 2019, using molecular methods.

Real-time PCR targeting the *lipL32* gene [27] or *rrs* (16S) gene [29] was used to screen the infected dogs and genotyping was conducted to identify the leptospires responsible for the infections.

The isolation of *Leptospira* strains was attempted, but only 15 out of 486 samples were successfully cultured, despite the immediate inoculation of the culture medium with samples. This confirmed that culturing *Leptospira* is difficult [22] and indicates the importance of increasing the tools available for genotyping strains directly from the DNA of biological samples.

The genotyping analysis, which was successfully performed on 15 isolates and 120 leptospiral DNAs, has increased our understanding of the epidemiological status of canine leptospirosis in Northeast Italy. It corroborated the serological evidence of the presence of high frequencies of strains from serogroup Icterohaemorrhagiae (ST17) and Australis (ST24 and ST198), but it also revealed the important roles of other serogroups, such as Pomona (ST117 and ST289) and Sejroe (ST155), as causative agents of the disease. Furthermore, this study revealed the presence of two recent genotypes, ST198 and ST289, which were discovered in hedgehogs and pigs, respectively, in Northern Italy. Thus, the results offer an interesting epidemiological perspective on the circulation of strains among different species and areas.

Among the identified genotypes, ST17, indicating the presence of *L. interrogans* serogroup Icterohaemorrhagiae, was confirmed as the major cause of canine leptospirosis. Previous serological surveys have consistently shown Icterohaemorrhagiae as the most reactive serogroup in dogs suspected of having leptospirosis [13,15,43–46] and interestingly, the serological profiles of human subjects suspected of having leptospiral infections also indicate that serovars Icterohaemorrhagiae and Copenhageni act as the main causative agents [47,48].

The application of an epidemiological approach, which considered data from the National Reference Centre for Animal Leptospirosis of Istituto Zooprofilattico Sperimentale della Lombardia e dell'Emilia Romagna (IZSLER) related to historical canine samples or belonging to other host species, provided insights into the infection's transmission chain. Our findings corroborated the hypothesis that dogs are exposed to environmental contamination spread by rodents [49], notably *Rattus* spp., which act as the main reservoir host of Icterohaemorrhagiae/Copenhageni serovars worldwide [50]. Although the inclusion of serogroup Icterohaemorrhagiae in vaccine formulations has led to its decreasing seroprevalence in Italy [46], this serogroup is still a major cause of disease in dogs, owing to the ubiquitous nature of its maintenance host. The dog, in this case, represents a spill-over and can be used as an important sentinel species for human and other animal infections [51–53]. Interestingly, the IZSLER database revealed that *Leptospira* ST17 occurs in different hosts (mouse, cat, hedgehog, horse, goat and cow), which indicates its ability to cause infection in a wide range of animal species. The sequencing of the *lic12008* gene, which was recently reported by Santos et al. [54] as useful in discriminating Icterohaemorrhagiae and Copenhageni serovars, allowed us to determine that the major portion (94%) of samples typed as ST17 represented serovar Icterohaemorrhagiae, and only a small portion (6%) represented serovar Copenhageni. The reliability of this molecular test for serovar discrimination is still unclear, and further studies are necessary, given that the serovar determinations of two isolates, as assessed by mAbs and by *lic12008* sequencing, were not in agreement (Table 5).

Three regularly vaccinated dogs were infected by *L. interrogans* serogroup Icterohaemorrhagiae, which is included in all the commercial vaccine formulations, which, however, are commonly able to protect the animal against acute signs but may not prevent infection if the animal is exposed to high bacterial load [15]. An evaluation of the anamnestic and prognostic data of these dogs revealed that they were borderline cases. The puppy, being in a vulnerable developmental period, showed symptoms even after it developed antibodies, but the protection provided by the vaccine may have played a key role in its complete recovery. The hunting dog, which may have been exposed to high bacterial loads that could be responsible for the disease manifestation, started therapy late, resulting in its death. In this case, the delayed treatment of leptospirosis may have allowed the escape of surviving bacteria to the bactericidal effect of antibiotics. The dog that commonly frequented the river and developed acute symptoms with severe hepatonephritis, responded to therapy and completely recovered, probably owing to an active and ready immune response previously stimulated by the vaccine. With the exception of the second case, in which the delay in receiving therapy may have been fatal, vaccination appears to have played a crucial role in protecting dogs from death, given that Icterohaemorrhagiae infections in dogs are often associated with clinical courses culminating in fatal outcomes [15,55,56].

Interestingly, 13 dogs were infected by *L. interrogans* ST198, a new genotype first revealed in hedgehogs of Northern Italy [41] and continually serologically typed as belonging to serovar Australis and serogroup Australis. Very recently, Balboni et al. [57] identified ST198 in a symptomatic dog, providing further evidence of its involvement in canine leptospirosis in Italy. The discovery of this genotype among dogs suggested two important findings: from an epidemiological point of view, a possible direct transmission between species or by indirect infection through a contaminated shared environment, and from a clinical point of view, its presence and ability to cause the clinical presentation of leptospirosis, which can be fatal in Italian dogs.

In this study, a novel genotype, called "ST17-like" because of its similarity to ST17, was discovered in a dog of 2019. Unfortunately, it was unsuitable for submission to BIGSdb because of its incomplete MLST profile. Further confirmation and investigations are required, including the evaluation of its presence in other dogs and hosts in the same area.

More than 7% of the genotyped samples were characterized as ST24 and found worldwide in isolates serologically typed as belonging to serovars Jalna and Bratislava, within serogroup Australis [58]. Unfortunately, because the isolates were unavailable, they were not characterized using mAbs, and consequently, there is no data that can be used to define the specific infective serovars present in the sampled dogs. It is probable that these infections were largely caused by *L. interrogans* serovar Bratislava, because data from the past 30 years indicate the widespread exposure of dogs to this serovar in Europe [11,12,46]. This was confirmed by the isolation of this serovar from urine of infected dogs in Scotland [59] and by its presence among hedgehogs [60], pigs [12] and wild boars [61] in Italy. Bratislava is the serovar most commonly isolated from domestic animals and, for this reason, is most commonly used as the antigen to represent the Australis serogroup in MAT panels. Our study revealed that this strain is responsible for infections in several hosts other than dog, such as hedgehog, wild boar and horse, in agreement with previous serological studies [12,13,61]. Interestingly, in this study, ST24 was also identified in a fox, indicating possible transmission through a contaminated shared environment.

The coverage of the vaccine for the Australis serogroup remains to be evaluated, not only because bivalent vaccines are still widely used but also because new emerging strains, such as *Leptospira* serovar Australis ST198, may represent a significant cause of canine leptospirosis in Italy. Because current tetravalent vaccines contain antigens of Australis Bratislava (ST24), new research is required to evaluate their ability to protect dogs against infection by *Leptospira* Australis ST198.

Strains typed as ST117, indicating the presence of *L. kirschneri* serovar Mozdok, play an important role in causing leptospirosis in dogs. Epidemiological studies have identified this serovar across Europe [17,62–65], in dogs [19], small rodents [65–67], in cattle and pigs [68]. It has been involved in human cases and in canine infections in Cuba [64,66,67,69], making it a relevant risk to public health. Clinically, dogs infected with *Leptospira* belonging to this serogroup experience a severe disease characterized by lethargy, fever, lack of appetite, diffuse hemorrhage and renal and liver failure [70,71]. Serovar Pomona, included in the same serogroup, has emerged in the USA as cause of clinical symptoms in dogs [51], and it has been added to vaccine formulations. In Italy, in 2002, Scanziani and colleagues reported some serological positivity of dogs to Pomona serogroup [11], and recently, Bertelloni found an increasing incidence of Pomona among dogs in the north central area of Italy [13]. In 2010, Ellis [17] had proposed the usefulness of including other serovars, such as Pomona, in the vaccine formulation, but he also believed that additional clinical, cultural and serological studies were needed to support their inclusion. Here, the obtained data agree with previous serological data and support the inclusion of a Pomona serogroup strain, especially one belonging to serovar Mozdok, in dog vaccines in Europe.

Interestingly, five dogs were infected by leptospires genotyped as ST289, a new genotype recently found for the first time by IZSLER laboratory in a pig of 2014, living in Northern Italy [42]. It was typed using core genome MLST as very similar, but not identical, to *L. kirschneri* serovar Mozdok, within serogroup Pomona. The discovery of this genotype in Italian dogs suggested its ability to cause clinical manifestations in dogs and the possible direct transmission among animals or an indirect passage, possibly in a herd context. More studies should be performed to clarify the role of this new strain in causing canine leptospirosis, but our findings support the inclusion of serogroup Pomona in vaccine formulations

In the present study, ST155 was found in six dogs. It characterizes *L. borgpetersenii* serogroup Sejroe, but the serovar status was not deducible from the MLST profile. Nevertheless, the international database BIGSdb [58] reported that isolates typed as ST155 belong to serovars Polonica and Saxkoebing. Using the information from our database, ST155 was compatible with both serovar Sejroe and serovar Polonica, but not with serovar Saxkoebing. In fact, the reference strains serovar Sejroe strain Topino 1, serovar Sejroe strain M24 (Figure 2) and serovar Polonica strain 493 Poland (not reported) were previously typed as ST155, while the reference strain serovar Saxkoebing strain Mus 24 was typed as ST219 (Figure 2). Strains serotyped as Sejroe were previously described as being causative agents of canine leptospirosis worldwide [72,73], and interestingly, this strain was found in an asymptomatic dog in Brazil [74,75]. The minimum spanning tree, color-coded by animal host, revealed the shared presence of these STs among dogs, horses and mice.

Our results indicated that the use of the serovar of the serogroup Sejroe strain in MAT analyses to test local canine populations should be reconsidered. In fact, the MAT panel commonly contains serovar Hardjo (ST152), but the inclusion of serovar Sejroe (ST155) would be more useful for diagnosing canine leptospirosis in our area.

Serogroup Canicola, historically included in commercial vaccine formulations, was not found in this work, and this was in agreement with previous serological studies that describe a decline in the prevalence of this serogroup in European countries, possibly because of the widespread protection resulting from the vaccination of the reservoir host [76], i.e., dog. The risk of stopping vaccinations against host-adapted *Leptospira* in its target host is that its prevalence may rapidly increase once the population's immunity falls; therefore, the vaccination of dogs against serovar Canicola should continue [17].

Serovar Grippotyphosa is maintained by a number of small rodent species in mainland Europe [77–79]. Seroprevalence studies have indicated Grippotyphosa infections in several European countries, including Italy [11], and consequently, it has been included in commercial vaccines. However, in this molecular survey, this strain was not detected among the sampled canine population nor in other species.

The geographical map indicated a widespread distribution of the identified genotypes throughout the considered area, except for ST289, mostly concentrated in Emilia-Romagna region and ST117, found only in Veneto region. Nevertheless, these differences might be attributed to the variability in the numbers of genotyped samples rather than a geographic-dependent diversity in ST circulation. Proper investigations will determine the presence of specific environmental niches for canine leptospirosis or particular risk factors, but the common practice of owners traveling with their pets for recreational activities, such as hiking and hunting, often out of their town or province, should be taken into consideration during risk evaluation. Therefore, when planning dog vaccinations or investigating a suspected clinical case, all of the identified genotypes should be considered epidemiologically relevant.

4. Materials and Methods

4.1. Sampling, DNA Extraction and Real-Time PCR

In total, 2485 biological samples from 1631 dogs with suspected clinical leptospirosis were tested at IZSLER located in Brescia (Italy) and at Istituto Zooprofilattico Sperimentale delle Venezie (IZSVE) located in Legnaro (Padua, Italy) between January 2013 and December 2019.

Samples were collected by veterinary practitioners in the north central area of Italy (Lombardy and Emilia-Romagna regions) and in Northeast Italy (Veneto, Friuli Venezia Giulia and Trentino- Alto Adige regions) during routine diagnostic activities, owing to a systematic passive survey funded by the Italian Ministry of Health and coordinated by the IZSVE, involving many veterinary practitioners of the region. When available, data on the vaccination states and clinical follow-ups of dogs were also recorded.

DNA extractions and real-time PCR targeting the *lipL32* gene were carried out at IZSLER as previously described by Bertasio et al. [42], while at IZSVE, different DNA extraction methods and kits were used in accordance with the variety of samples. A commercially available High Pure PCR Template Preparation kit (Roche Diagnostics, Mannheim, Germany) was used, in accordance with the manufacturer's instructions, to extract 2 mL of urine or 1 cm^3 tissue homogenate. From 1 mL of EDTA-treated blood, DNA was extracted using a commercially available QIAamp DNA Mini kit (Qiagen, Hilden, Germany), in accordance with the manufacturer's instructions. In the case of poor-cellular matrices, as in urine and blood samples, 20 µg of a poly-A carrier (Roche, Diagnostics, Mannheim, Germany) was added to each sample to increase the recovery efficiency of nucleic acids. All the DNA extraction preparations included a negative control (water). All the DNAs extracted from biological samples were subjected to a TaqMan-based real-time PCR assay targeting an 87-bp fragment that corresponded to a portion of the gene encoding the 16S rDNA [29].

The PCR was performed in a 25 µL final volume, containing 3 µL of extracted DNA, 12. 5 µL of 2× Master Mix TaqMan Universal 2× (Thermo Fisher Scientific, Waltham, MA, USA), 300 nM of each primer and 100 nM of a 5′ FAM–3′-TAMRA probe. All the amplification assays included a negative control (water), a negative bacterial genomic control (DNA of *Leptospira biflexa* serovar Patoc) and a positive control (DNA of *L. interrogans* serovar Icterohaemorrhagiae), and each sample was tested in duplicate. The assay was performed on a 7900HT Fast Real-time PCR System (Thermo Fisher Scientific) with the following thermal conditions: a hot-start step at 50 °C for 2 min, a holding step at 95 °C for 10 min and 45 cycles of 95 °C for 15 s and 60 °C for 60 s. Samples with $C_t < 38$ were considered positive. Samples having C_t values within the 38–40 range were considered doubtful, whereas samples having no FAM fluorescence signal or with $C_t \geq 40$ were considered negative.

The two methods were validated using reference materials and their performance monitored by a regular participation to proficiency testing programs organized by the National Reference Laboratory, with an agreement index (K Cohen) equal to 1.

4.2. Isolation and Serological Typing of Isolates

In total, 486 samples from 449 clinically affected dogs were chosen to attempt the isolation of the causal agent as previously described [42]. The specific matrices of these samples were 416 urines, 35 kidney and 11 livers, while for 24 samples, the data were unavailable.

The preliminary serogroup assignment was carried out using MAT with polyclonal serogroup-specific antibodies, as previously described [42]. The serovars were classified using MAT with different panels of mAbs for Icterohaemorrhagiae, Pomona and Australis serogroups. The mAbs used were the following: five mAbs (F20 C4, F52 C1, F70 C24, F70 C26 and F82 C1) for the Icterohaemorrhagiae serogroup [80], five mAbs (F43 C9, F46 C9, F48 C6, F58 C1 and F61 C7) for the Pomona serogroup [81] and six mAbs (F81 C3, F81 C5, F81 C6, F81 C8, F90 C6 and F90 C8) for the Australis serogroup [82]. These mAbs and the polyclonal anti-sera were previously purchased from the OIE Leptospirosis Reference Centre, Royal Tropical Institute (KIT), (Amsterdam, The Netherlands), and the protocol was in accordance with the standard serological methods used in this reference laboratory. To interpret the results, we referred to the expected maximum dilution titers provided by the Leptospirosis Reference Centre of KIT.

4.3. Genotyping

4.3.1. MLST

Until 2015, MLST analyses were applied only to DNA samples with $C_t \leq 35$ as assessed by real-time PCR, and starting from 2016, the use of MLST was extended to all DNA samples. If more than one biological matrix was available for the same dog, the analysis was attempted on all the samples to maximize the probability of success. In total, 223 samples (15 isolates and 208 leptospiral DNAs) belonging to 193 dogs were submitted for MLST analysis (Table 6).

Table 6. Samples submitted for MLST analysis: details of their matrices.

Matrix	No. of DNA Extracted from Biological Samples	No. of Isolated Samples	Total
Urine	101	15	120
Blood	45		45
Kidney	34		34
Lung	5		5
Organ(s)	6		6
Liver	4		4
Bladder	3		3
Not defined	10		10
Total	208	15	223

The analysis was performed using the 7-loci scheme proposed by Boonsilp in 2013 [36], which is based on the housekeeping genes *glmU*, *pntA*, *sucA*, *tpiA*, *pfkB*, *mreA* and *caiB*, as previously described [42]. Nucleotide sequences were assembled using the SeqMan module of the Lasergene sequencing analysis software package (DNASTAR Inc., Madison, WI, USA) or using Bionumerics software ver. 7.6 (Applied Math, Biomerieux, Sint-Martens-Latem, Belgium). For allelic number and ST identification, assembled and trimmed sequences were queried against the BIGSdb available on the *Leptospira* MLST website (https://pubmlst.org/leptospira/) sited at the University of Oxford [58] or by an automated analysis using Bionumerics. Comparisons between the STs found and those present in BIGSdb as reference isolates were used to deduce the species of the *Leptospira* being tested. To perform comparisons among historical serological studies (where serovars and serogroups were defined) and genotyping data (where species and genomic profiles were defined), we chose to assign to each identified ST a classification at the serogroup and serovar levels obtained from BIGSdb, knowing that this information was deduced and did not result from active serological typing.

4.3.2. MLVA

A VNTR analysis was performed on 21 samples (15 isolates and 6 DNAs). Five discriminatory loci (VNTR-4, -7, -10, Lb4 and Lb5) described by Salaün and colleagues in 2006 [37] were considered, as previously described [42].

4.4. Sequencing of a Tract of the lic12008 Gene

A tract of the *lic12008* gene of samples genotyped as ST17 (characterized as *L. interrogans* serogroup Icterohaemorrhagiae) was sequenced. The tract was recently found to be involved in the genetic distinction between serovars Icterohaemorrhagiae and Copenhageni [54], which are indistinguishable by MLST and MLVA. The analysis determines the presence of a single base insertion of a thymine nucleotide within a poly-thymine tract (9-bp long) in the *lic12008* of all *L. interrogans* serovar Icterohaemorrhagiae isolates but not in Copenhageni strains. We used gene-specific primers (forward 5′-TAGGTTGGCACGAAGGTTCT-3′ and reverse 5′-CTTAAACTTTCCACTTTCCGGA-3′) to amplify a short 163-bp genomic tract containing the described discriminatory INDEL.

The PCR amplification was performed using a KAPA2G Robust HotStart PCR kit (Kapabiosystems Resnova, Rome, Italy) in a 25 μL total volume containing 0.4 μM each primer and 5 μL of DNA. The thermal conditions were as follows: 1 cycle at 95 °C for 7 min, 45 cycles at 95 °C for 30 s, 53 °C for 30 s and 72 °C for 30 s, followed by a final elongation at 72 °C for 5 min. For the sequencing procedure, we followed the same protocol used for the MLST analysis. Nucleotide sequences were assembled using the SeqMan module of the Lasergene sequencing analysis software package (DNASTAR) and were aligned with the sequences of Copenhageni and an Icterohaemorrhagiae reference strains to permit a comparison and to define the serovar of the tested *Leptospira* strain.

4.5. Analysis of Data

4.5.1. Geographical Map

The provinces in which the dogs were living were recorded to construct a map that displayed the distributions of the identified genotypes. The map was produced using the commercial software ESRI^{TM}ArcMap 10.5.1. The spatial layer containing the provinces was joined (using province ID attribute) with the sample data. Specific style roles were applied to generate different pie charts based on the presence of different STs. The size of each chart represents the total number of genotyped samples per province. Each chart reports the results (as percentages) per ST (denoted by different colors).

4.5.2. Phylogenetic Analysis of Concatenated MLST Loci

The concatenated loci assessed by MLST on our samples were compared with those of reference strains purchased from the Leptospirosis Reference Centre of KIT using the UPGMA method with a bootstrap of 1000 replicates by Bionumerics ver. 7.6 (Applied Maths, Sint-Martens-Latem, Belgium). The resulting tree reported sample details, such as the identification number, the species, the serological classification (real for isolates and presumptive for DNA samples) and the MLVA pattern.

4.5.3. Minimum Spanning Tree based on STs

An analysis of the characteristics of alleles assessed by MLST was performed by the creation of a minimum spanning tree using Bionumerics software ver. 7.6 (Applied Maths). The analysis was applied only to samples having a complete MLST profile, and to perform an epidemiological evaluation, we also considered data for some samples already present in the IZSLER database but external to the period included in this paper (six isolated strains from dogs: 2, 1, 1, 1 and 1 from 1992, 1995, 1996, 1999 and 2007, respectively) or belonging to other hosts species (20 from rat, 3 from mouse, 2 from cat, 14 from hedgehog, 3 from horse, 1 from cow, 1 from goat, 1 from pig, 1 from wolf and 1 from wild boar). The IZSLER database is an internal database file that includes updated information on the genotyping and serological typing of isolated strains preserved in liquid nitrogen, on field samples from various animal species collected from 1989 and on reference strains purchased from by the Leptospirosis Reference Centre of KIT. Furthermore, starting in 2016, data from the molecular typing of leptospiral DNAs extracted directly from biological samples of a series of animal hosts have been added.

5. Conclusions

To the best of our knowledge this is the first study that describes the genetic diversity of pathogenic *Leptospira* in dogs in Italy, and it has increased our understanding of the related epidemiological situation among the Italian canine population. No area of specific risk was revealed in Northeast Italy, and the widespread presence of already known infecting serogroups, like Icterohaemorrhagiae and Australis, was confirmed. Interestingly, it highlighted the circulation of infecting strains belonging to other serogroups, such as Pomona and Sejroe, that are not included in the commercial vaccines currently available in Italy. The strains of these two serogroups were responsible for clinical leptospirosis, which was sometimes fatal, in dogs. Furthermore, new strains recently observed in other hosts in

Italy, like ST198 from hedgehogs and ST289 from pigs, may play important roles in causing canine leptospirosis in Italy.

Author Contributions: Conceptualization, A.N., M.B.B., M.D. and L.L.; methodology, C.B., L.B. and M.M.; software, C.B. and M.M.; formal analysis, C.B., L.L. and L.C.; investigation, A.N., L.L. and T.F.; resources, A.N., L.L., M.B.B., M.D. and T.F; data curation, C.B. and L.L.; writing—original draft preparation, C.B.; writing—review and editing, M.M.B., L.L., L.C., T.F., M.D. and A.N.; supervision, M.B.B. and A.N.; project administration, A.N. and M.B.B.; funding acquisition, M.B.B., M.D. and A.N. All authors have read and agreed to the published version of the manuscript.

Funding: This research was funded by the Italian Ministry of Health, grant number IZSVE RC 16/12, IZSVE RC 05/2017 and IZSLER PRC2017016.

Acknowledgments: We thank Arturo Scalvenzi, Marcello Fin, Eufrasia Peroni Ammaturo, Cosima Tomasoni, Silvia Marchione and Isabella Giurisato for their skilled technical assistance with the bacterial isolation and serological typing.

Conflicts of Interest: The authors declare no conflict of interest. The funders had no role in the design of the study; in the collection, analyses, or interpretation of data; in the writing of the manuscript or in the decision to publish the results.

References

1. Schuller, S.; Francey, T.; Hartmann, K.; Hugonnard, M.; Kohn, B.; Nally, J.E.; Sykes, J. European consensus statement on leptospirosis in dogs and cats. *J. Small Anim. Pract.* **2015**, *56*, 159–179. [CrossRef] [PubMed]
2. Klarenbeek, A.; Schuffner, W.A.P. Appearance in Holland of Leptospira differing from Weil Strain. *Ned. Tijdschr. Geneeskd.* **1933**, *77*, 4271–4276.
3. Azócar-Aedo, L.; Monti, G. Meta-Analyses of Factors Associated with Leptospirosis in Domestic Dogs. *Zoonoses Public Health* **2016**, *63*, 328–336. [CrossRef] [PubMed]
4. Zwijnenberg, R.J.G.; Smythe, L.D.; Symonds, M.I.; Dohnt, M.F.; Toribio, J.L.M.L. Cross-sectional study of canine leptospirosis in animal shelter populations in mainland Australia. *Aust. Vet. J.* **2008**, *86*, 317–323. [CrossRef] [PubMed]
5. Shi, D.; Liu, M.; Guo, S.; Liao, S.; Sun, M.; Liu, J.; Wang, L.; Wang, Z.; Wang, S.; Yang, D.; et al. Serological survey of canine leptospirosis in Southern China. *Pak. Vet. J.* **2012**, *32*, 280–282.
6. De Abreu, J.A.P.; da Silva Krawczak, F.; Guedes, I.B.; de Souza-Filho, A.F.; de Souza, G.O.; de Campos Binder, L.; de Oliveira, C.S.; Sponchiado, J.; Melo, G.L.; Labruna, M.B.; et al. Frequency of anti-Leptospira spp. antibodies in dogs and wild small mammals from rural properties and conservation units in southern Brazil. *One Health* **2019**, *8*, 100104. [CrossRef]
7. Khalili, M.; Sakhaee, E.; Amiri, F.B.; Safat, A.A.; Afshar, D.; Esmaeili, S. Serological evidence of leptospirosis in Iran; A systematic review and meta-analysis. *Microb. Pathog.* **2020**, *138*, 103833. [CrossRef]
8. Davis, M.A.; Evermann, J.F.; Petersen, C.R.; VancerSchalie, J.; Besser, T.E.; Huckabee, J.; Daniels, J.B.; Hancock, D.D.; Leslie, M.; Baer, R. Serological survey for antibodies to Leptospira in dogs and raccoons in Washington State. *Zoonoses Public Health* **2008**, *55*, 436–442. [CrossRef]
9. López, M.C.; Vila, A.; Rodón, J.; Roura, X. Leptospira seroprevalence in owned dogs from Spain. *Heliyon* **2019**, *5*, e02373. [CrossRef]
10. Ambily, R.; Mini, M.; Joseph, S.; Krishna, S.V.; Abhinai, G. Canine leptospirosis—A seroprevalence study from Kerala, India. *Vet. World* **2013**, *6*, 42–44. [CrossRef]
11. Scanziani, E.; Origgi, F.; Giusti, A.M.; Iacchia, G.; Vasino, A.; Pirovano, G.; Scarpa, P.; Tagliabue, S. Serological survey of leptospiral infection in kennelled dogs in Italy. *J. Small Anim. Pract.* **2002**, *43*, 154–157. [CrossRef] [PubMed]
12. Tagliabue, S.; Figarolli, B.M.; D'Incau, M.; Foschi, G.; Gennero, M.S.; Giordani, R.; Natale, A.; Papa, P.; Ponti, N.; Scaltrito, D.; et al. Serological surveillance of Leptospirosis in Italy: Two-year national data (2010–2011). *Vet. Ital.* **2016**, *52*, 129–138. [CrossRef] [PubMed]
13. Bertelloni, F.; Cilia, G.; Turchi, B.; Pinzauti, P.; Cerri, D.; Fratini, F. Epidemiology of leptospirosis in North-Central Italy: Fifteen years of serological data (2002–2016). *Comp. Immunol. Microbiol. Infect. Dis.* **2019**, *65*, 14–22. [CrossRef] [PubMed]
14. Jull, D.J.; Heath, K.R. The Evaluation of a Combined L.canicola and L.icterohaemorrhagiae Vaccine on Hamsters and Dogs. *J. Small Anim. Pract.* **1960**, *1*, 245–258. [CrossRef]

15. André-Fontaine, G. Canine leptospirosis-Do we have a problem? *Vet. Microbiol.* **2006**, *117*, 19–24. [CrossRef]
16. Sykes, J.E.; Hartmann, K.; Lunn, K.F.; Moore, G.E.; Stoddard, R.A.; Goldstein, R.E. 2010 ACVIM Small Animal Consensus Statement on Leptospirosis: Diagnosis, Epidemiology, Treatment, and Prevention. *J. Vet. Intern. Med.* **2011**, *25*, 1–13. [CrossRef]
17. Ellis, W.A. Control of canine leptospirosis in Europe: Time for a change? *Vet. Rec.* **2010**, *167*, 602–605. [CrossRef]
18. Mayer-Scholl, A.; Luge, E.; Draeger, A.; Nöckler, K.; Kohn, B. Distribution of Leptospira Serogroups in Dogs from Berlin, Germany. *Vector-Borne Zoonotic Dis.* **2013**, *13*, 200–202. [CrossRef]
19. Renaud, C.; Andrews, S.; Djelouadji, Z.; Lecheval, S.; Corrao-Revol, N.; Buff, S.; Demont, P.; Kodjo, A. Prevalence of the Leptospira serovars bratislava, grippotyphosa, mozdok and pomona in French dogs. *Vet. J.* **2013**, *196*, 126–127. [CrossRef]
20. Klaasen, H.L.B.M.; Van Der Veen, M.; Molkenboer, M.J.C.H.; Sutton, D. A novel tetravalent Leptospira bacterin protects against infection and shedding following challenge in dogs. *Vet. Rec.* **2013**, *172*, 181. [CrossRef]
21. Klaasen, H.L.B.M.; van der Veen, M.; Sutton, D.; Molkenboer, M.J.C.H. A new tetravalent canine leptospirosis vaccine provides at least 12 months immunity against infection. *Vet. Immunol. Immunopathol.* **2014**, *158*, 26–29. [CrossRef] [PubMed]
22. Levett, P.N. Leptospirosis. *Clin. Microbiol. Rev.* **2001**, *14*, 296–326. [CrossRef]
23. Leptospirosis. In *OIE Terrestrial Manual*; World Organization for Animal Health: Paris, France, 2018; pp. 503–516.
24. Strutzberg-Minder, K.; Tschentscher, A.; Beyerbach, M.; Homuth, M.; Kreienbrock, L. Passive surveillance of Leptospira infection in swine in Germany. *Porc. Health Manag.* **2018**, *4*, 10. [CrossRef] [PubMed]
25. Levett, P.N. Usefulness of Serologic Analysis as a Predictor of the Infecting Serovar in Patients with Severe Leptospirosis. *Clin. Infect. Dis.* **2003**, *36*, 447–452. [CrossRef] [PubMed]
26. Martin, L.E.R.; Wiggans, K.T.; Wennogle, S.A.; Curtis, K.; Chandrashekar, R.; Lappin, M.R. Vaccine-Associated *Leptospira* Antibodies in Client-Owned Dogs. *J. Vet. Intern. Med.* **2014**, *28*, 789–792. [CrossRef]
27. Stoddard, R.A.; Gee, J.E.; Wilkins, P.P.; McCaustland, K.; Hoffmaster, A.R. Detection of pathogenic Leptospira spp. through TaqMan polymerase chain reaction targeting the LipL32 gene. *Diagn. Microbiol. Infect. Dis.* **2009**, *64*, 247–255. [CrossRef]
28. Levett, P.N.; Morey, R.E.; Galloway, R.L.; Turner, D.E.; Steigerwalt, A.G.; Mayer, L.W. Detection of pathogenic leptospires by real-time quantitative PCR. *J. Med. Microbiol.* **2005**, *54*, 45–49. [CrossRef]
29. Smythe, L.D.; Smith, I.L.; Smith, G.A.; Dohnt, M.F.; Symonds, M.L.; Barnett, L.J.; McKay, D.B. A quantitative PCR (TaqMan) assay for pathogenic *Leptospira* spp. *BMC Infect. Dis.* **2002**, *2*, 13. [CrossRef]
30. Merien, F.; Portnoi, D.; Bourhy, P.; Charavay, F.; Berlioz-Arthaud, A.; Baranton, G. A rapid and quantitative method for the detection of Leptospira species in human leptospirosis. *FEMS Microbiol. Lett.* **2005**, *249*, 139–147. [CrossRef]
31. Ahmed, A.; Engelberts, M.F.M.; Boer, K.R.; Ahmed, N.; Hartskeerl, R.A. Development and validation of a real-time PCR for detection of pathogenic Leptospira species in clinical materials. *PLoS ONE* **2009**, *4*, e7093. [CrossRef]
32. Slack, A.T.; Symonds, M.L.; Dohnt, M.F.; Smythe, L.D. Identification of pathogenic Leptospira species by conventional or real-time PCR and sequencing of the DNA gyrase subunit B encoding gene. *BMC Microbiol.* **2006**, *6*, 95. [CrossRef] [PubMed]
33. Bourhy, P.; Bremont, S.; Zinini, F.; Giry, C.; Picardeau, M. Comparison of real-time PCR assays for detection of pathogenic Leptospira spp. in blood and identification of variations in target sequences. *J. Clin. Microbiol.* **2011**, *49*, 2154–2160. [CrossRef] [PubMed]
34. Harkin, K.R.; Roshto, Y.M.; Sullivan, J.T.; Purvis, T.J.; Chengappa, M.M. Comparison of polymerase chain reaction assay, bacteriologic culture, and serologic testing in assessment of prevalence of urinary shedding of leptospires in dogs. *J. Am. Vet. Med. Assoc.* **2003**, *222*, 1230–1233. [CrossRef]
35. Midence, J.N.; Leutenegger, C.M.; Chandler, A.M.; Goldstein, R.E. Effects of recent Leptospira vaccination on whole blood real-time PCR testing in healthy client-owned dogs. *J. Vet. Intern. Med.* **2012**, *26*, 149–152. [CrossRef] [PubMed]

36. Boonsilp, S.; Thaipadungpanit, J.; Amornchai, P.; Wuthiekanun, V.; Bailey, M.S.; Holden, M.T.G.; Zhang, C.; Jiang, X.; Koizumi, N.; Taylor, K.; et al. A single multilocus sequence typing (MLST) scheme for seven pathogenic Leptospira species. *PLoS Negl. Trop. Dis.* **2013**, *7*, e1954. [CrossRef] [PubMed]
37. Salaün, L.; Mérien, F.; Gurianova, S.; Baranton, G.; Picardeau, M. Application of Multilocus Variable-Number Tandem-Repeat Analysis for Molecular Typing of the Agent of Leptospirosis. *J. Clin. Microbiol.* **2006**, *44*, 3954–3962. [CrossRef]
38. Weiss, S.; Menezes, A.; Woods, K.; Chanthongthip, A.; Dittrich, S.; Opoku-Boateng, A.; Simuli, M.; Chalke, V. An Extended Multilocus Sequence Typing (MLST) Scheme for Rapid Direct Typing of Leptospira from Clinical Samples. *PLoS Negl. Trop. Dis.* **2016**, *10*, e0004996. [CrossRef]
39. Klaasen, H.L.B.M.; Molkenboer, M.J.C.H.; Vrijenhoek, M.P.; Kaashoek, M.J. Duration of immunity in dogs vaccinated against leptospirosis with a bivalent inactivated vaccine. *Vet. Microbiol.* **2003**, *95*, 121–132. [CrossRef]
40. Wilson, S.; Stirling, C.; Thomas, A.; King, V.; Plevová, E.; Chromá, L.; Siedek, E.; Illambas, J.; Salt, J.; Sture, G. Duration of immunity of a multivalent (DHPPi/L4R) canine vaccine against four Leptospira serovars. *Vaccine* **2013**, *31*, 3126–3130. [CrossRef]
41. Boniotti, M.B.; Gelmini, L.; Carra, E.; Figarolli, B.M.; D'incau, M.; Tagliabue, S. Leptospira Interrogans Serogroup Australis In Hedgehog In Northern Italy. Proocedings of the International Leptospirosis Society of the Conference, Fukuoka, Japan, 7–11 October 2013.
42. Bertasio, C.; Papetti, A.; Scaltriti, E.; Tagliabue, S.; D'Incau, M.; Boniotti, M.B. Serological Survey and Molecular Typing Reveal New Leptospira Serogroup Pomona Strains among Pigs of Northern Italy. *Pathogens* **2020**, *9*, 332. [CrossRef]
43. Modric, Z.; Culjak, K.; Hahn, V. Leptospirosis in a dog caused by Leptospira interrogans serotype pomona. *Vet. Glas.* **1987**, *41*, 43–47.
44. Vojinović, D.; Bogićević, N.; Vasić, A.; Manić, M.; Elezović Radovanović, M.; Rogožarski, D.; Marić, J.; Valčić, M. Seroepidemiological survey of leptospiral infection in stray dogs in Serbia. *Turkish J. Vet. Anim. Sci.* **2015**, *39*, 719–723. [CrossRef]
45. Burriel, A.R.; Dalley, C.; Woodward, M.J. Prevalence of Leptospira species among farmed and domestic animals in Greece. *Vet. Rec.* **2003**, *153*, 146–148. [CrossRef]
46. Cerri, D.; Ebani, V.V.; Fratini, F.; Pinzauti, P.; Andreani, E. Epidemiology of leptospirosis: Observations on serological data obtained by a "diagnostic laboratory for leptospirosis" from 1995 to 2001. *New Microbiol.* **2003**, *26*, 383–389.
47. Felzemburgh, R.D.M.; Ribeiro, G.S.; Costa, F.; Reis, R.B.; Hagan, J.E.; Melendez, A.X.T.O.; Fraga, D.; Santana, F.S.; Mohr, S.; dos Santos, B.L.; et al. Prospective Study of Leptospirosis Transmission in an Urban Slum Community: Role of Poor Environment in Repeated Exposures to the Leptospira Agent. *PLoS Negl. Trop. Dis.* **2014**, *8*, e2927. [CrossRef]
48. Oliveira, M.A.A.; Leal, É.A.; Correia, M.A.; Serufo Filho, J.C.; Dias, R.S.; Serufo, J.C. Human leptospirosis: Occurrence of serovars of Leptospira spp. in the state of Minas Gerais, Brazil, from 2008 to 2012. *Braz. J. Microbiol.* **2017**, *48*, 483–488. [CrossRef] [PubMed]
49. Suepaul, S.M.; Carrington, C.V.F.; Campbell, M.; Borde, G.; Adesiyun, A.A. Serovars of Leptospira isolated from dogs and rodents. *Epidemiol. Infect.* **2010**, *138*, 1059–1070. [CrossRef] [PubMed]
50. Boey, K.; Shiokawa, K.; Rajeev, S. Leptospira infection in rats: A literature review of global prevalence and distribution. *PLoS Negl. Trop. Dis.* **2019**, *13*, e0007499. [CrossRef]
51. Ghneim, G.S.; Viers, J.H.; Chomel, B.B.; Kass, P.H.; Descollonges, D.A.; Johnson, M.L. Use of a case-control study and geographic information systems to determine environmental and demographic risk factors for canine leptospirosis. *Vet. Res.* **2007**, *38*, 37–50. [CrossRef]
52. Major, A.; Schweighauser, A.; Francey, T. Increasing incidence of canine leptospirosis in Switzerland. *Int. J. Environ. Res. Public Health* **2014**, *11*, 7242–7260. [CrossRef]
53. Schuller, S.; Arent, Z.J.; Gilmore, C.; Nally, J. Prevalence of antileptospiral serum antibodies in dogs in Ireland. *Vet. Rec.* **2015**, *177*, 126. [CrossRef] [PubMed]
54. Santos, L.A.; Adhikarla, H.; Yan, X.; Wang, Z.; Fouts, D.E.; Vinetz, J.M.; Alcantara, L.C.J.; Hartskeerl, R.A.; Goris, M.G.A.; Picardeau, M.; et al. Genomic comparison among global isolates of L. interrogans serovars Copenhageni and Icterohaemorrhagiae identified natural genetic variation caused by an indel. *Front. Cell. Infect. Microbiol.* **2018**, *8*, 193. [CrossRef] [PubMed]

55. Rissi, D.R.; Brown, C.A. Diagnostic features in 10 naturally occurring cases of acute fatal canine leptospirosis. *J. Vet. Diagn. Investig.* **2014**, *26*, 799–804. [CrossRef] [PubMed]
56. CE, N.; GJ, K.; JJ, K. Serum biochemical changes in dogs with experimental Leptospira interrogans serovar icterohaemorrhagiae infection. *Am. J. Vet. Res.* **1981**, *42*, 1125–1129.
57. Balboni, A.; Zamagni, S.; Bertasio, C.; Boniotti, M.B.; Troìa, R.; Battilani, M.; Dondi, F. Identification of Serogroups Australis and Icterohaemorrhagiae in Two Dogs with a Severe Form of Acute Leptospirosis in Italy. *Pathogens* **2020**, *9*, 351. [CrossRef]
58. Jolley, K.A.; Maiden, M.C.J. BIGSdb: Scalable analysis of bacterial genome variation at the population level. *BMC Bioinform.* **2010**, *11*, 595. [CrossRef]
59. Van den Broek, A.H.M.; Thrusfield, M.V.; Dobbiet, G.R.; Ellisi, W.A. A serological and bacteriological survey of leptospiral infection in dogs in Edinburgh and Glasgow. *J. Small Anim. Pract.* **1991**, *32*, 118–124. [CrossRef]
60. Tagliabue, S.; Farina, R. Inchiesta sieroepidemiologica sulla diffusione della leptospirosi ta gli animali domestici ed alcune specie selvatiche. *Sel. Vet.* **1995**, *36*, 11–12.
61. Chiari, M.; Figarolli, B.M.; Tagliabue, S.; Alborali, G.L.; Bertoletti, M.; Papetti, A.; D'Incau, M.; Zanoni, M.; Boniotti, M.B. Seroprevalence and risk factors of leptospirosis in wild boars (Sus scrofa) in northern Italy. *Hystrix* **2016**, *27*, 2. [CrossRef]
62. Das Neves Paiva-Cardoso, M.; Arent, Z.; Gilmore, C.; Hartskeerl, R.; Ellis, W.A. Altodouro, a new Leptospira serovar of the Pomona serogroup isolated from rodents in northern Portugal. *Infect. Genet. Evol.* **2013**, *13*, 211–217. [CrossRef]
63. Rocha, T.; Ellis, W.A.; Montgomery, J.; Gilmore, C.; Regalla, J.; Brem, S. Microbiological and serological study of leptospirosis in horses at slaughter: First isolations. *Res. Vet. Sci.* **2004**, *76*, 199–202. [CrossRef]
64. Rocha, T. Isolation of Leptospira interrogans serovar mozdok from aborted swine fetuses in Portugal. *Vet. Rec.* **1990**, *126*, 602.
65. Majetic, Z.S.; Galloway, R.; Sabljic, E.R.; Milas, Z.; Perko, V.M.; Habus, J.; Margaletic, J.; Pernar, R.; Turk, N. Epizootiological survey of small mammals as Leptospira spp. reservoirs in Eastern Croatia. *Acta Trop.* **2014**, *131*, 111–116. [CrossRef] [PubMed]
66. Pritchard, D.G.; Todd, N.; Barlow, A.; Little, S.A. Outbreak of Leptospira interrogans serovar mozdok in sows in Dorset, England. *Isr. J. Vet. Med.* **1987**, *43*, 343.
67. Barlow, A.M. Reproductive failure in sows associated with Leptospira Mozdok from a wildlife source. *Pig J.* **2004**, *54*, 123–131.
68. Hathaway, S.C.; Todd, J.N.; Headlam, S.A.; Jeffrey, M. Possible role of leptospires of the Pomona serogroup in sporadic bovine abortion in the south west of England. *Vet. Rec.* **1984**, *115*, 623–626. [CrossRef]
69. Da Cunha, C.E.P.; Felix, S.R.; Neto, A.C.P.S.; Campello-Felix, A.; Kremer, F.S.; Monte, L.G.; Amaral, M.G.; Nobre, M.D.O.; Da Silva, É.F.; Hartleben, C.P.; et al. Infection with Leptospira kirschneri Serovar Mozdok: First Report from the Southern Hemisphere. *Am. J. Trop. Med. Hyg.* **2016**, *94*, 519–521. [CrossRef] [PubMed]
70. Greenlee, J.J.; Alt, D.P.; Bolin, C.A.; Zuerner, R.L.; Andreasen, C.B. Experimental canine leptospirosis caused by Leptospira interrogans serovars pomona and bratislava. *Am. J. Vet. Res.* **2005**, *66*, 1816–1822. [CrossRef] [PubMed]
71. Goldstein, R.E.; Lin, R.C.; Langston, C.E.; Scrivani, P.V.; Erb, H.N.; Barr, S.C. Influence of infecting serogroup on clinical features of leptospirosis in dogs. *J. Vet. Intern. Med.* **2006**, *20*, 489–494. [CrossRef] [PubMed]
72. Rühl-Fehlert, C.I.; Brem, S.; Feller, W.; Kopp, H.; Meyer, P.; Rinke, M. Clinical, microbiological and pathological observations in laboratory beagle dogs infected with leptospires of the serogroup sejroe. *Exp. Toxicol. Pathol.* **2000**, *52*, 201–207. [CrossRef]
73. Scanziani, E.; Crippa, L.; Giusti, A.M.; Luini, M.; Pacciarini, M.L.; Tagliabue, S.; Cavalletti, E. Leptospira interrogans serovar sejroe infection in a group of laboratory dogs. *Lab. Anim.* **1995**, *29*, 300–306. [CrossRef] [PubMed]
74. Miotto, B.A.; Tozzi, B.F.; de Souza Penteado, M.; Guilloux, A.G.A.; Moreno, L.Z.; Heinemann, M.B.; Moreno, A.M.; Lilenbaum, W.; Hagiwara, M.K. Diagnosis of acute canine leptospirosis using multiple laboratory tests and characterization of the isolated strains. *BMC Vet. Res.* **2018**, *14*, 222. [CrossRef]
75. Miotto, B.A.; Moreno, L.Z.; Guilloux, A.G.A.; de Sousa, G.O.; Loureiro, A.P.; Moreno, A.M.; Lilenbaum, W.; Vasconcellos, S.A.; Heinemann, M.B.; Hagiwara, M.K. Molecular and serological characterization of the first Leptospira santarosai strain isolated from a dog. *Acta Trop.* **2016**, *162*, 1–4. [CrossRef]

76. Claus, A.; Van de Maele, I.; Pasmans, F.; Gommeren, K.; Daminet, S. Leptospirosis in dogs: A retrospective study of seven clinical cases in Belgium. *Vlaams Diergeneeskd. Tijdschr.* **2008**, *77*, 259–263.
77. Fennestad, K.L.; Borg-Petersen, C. Leptospirosis in Danish wild mammals. *J. Wildl. Dis.* **1972**, *8*, 343–351. [CrossRef] [PubMed]
78. Borčić, B.; Kovačić, H.; Šebek, Z.; Aleraj, B.; Tvrtković, N. Small terrestrial mammals as reservoirs of leptospires in Sava valley (Croatia). *Folia Parasitol.* **2005**, *29*, 177–182.
79. Steinen, A.C.; Schuurman, J.L.; Gravekamp, C.; Korver, H.; Terpstra, W.J. Muskrats as carriers of pathogenic leptospires in The Netherlands. *Antonie Van Leeuwenhoek* **1992**, *61*, 43–50. [CrossRef]
80. Korver, H.; Kolk, A.; Vingerhoed, J.; Van Leeuwen, J.; Terpstra, W. Classification of serovars of the icterohaemorrhagiae serogroup by monoclonal antibodies. *Isr. J. Vet.* **1988**, *44*, 15–18.
81. Terpstra, W.J.; Korver, H.; Schoone, G.J.; von Leeuwen, J.; Schönemann, C.E.; de Jonge-Aglibut, S.; Kolk, A.H. Comparative classification of Leptospira serovars of the Pomona group by monoclonal antibodies and restriction-endonuclease analysis. *Zentralbl. Bakteriol. Mikrobiol. Hyg. A.* **1987**, *266*, 412–421. [CrossRef]
82. Cinco, M.; Banfi, E.; Stornello, C.; Campo, F.; Korver, H. First human isolate of Leptospira interrogans as serovar bratislava in Italy. *FEMS Microbiol. Lett.* **1989**, *47*, 499–503. [CrossRef]

© 2020 by the authors. Licensee MDPI, Basel, Switzerland. This article is an open access article distributed under the terms and conditions of the Creative Commons Attribution (CC BY) license (http://creativecommons.org/licenses/by/4.0/).

MDPI
St. Alban-Anlage 66
4052 Basel
Switzerland
Tel. +41 61 683 77 34
Fax +41 61 302 89 18
www.mdpi.com

Pathogens Editorial Office
E-mail: pathogens@mdpi.com
www.mdpi.com/journal/pathogens

www.ingramcontent.com/pod-product-compliance
Lightning Source LLC
Chambersburg PA
CBHW062101220926
51850CB00010B/2850
* 9 7 8 3 0 3 9 4 3 2 6 2 2 *